T0288873

For more information on this publication, visit www.rand.org/t/RR1755

**Library of Congress Cataloging-in-Publication Data** is available for this publication.
ISBN: 978-0-8330-9714-9

Published by the RAND Corporation, Santa Monica, Calif.
© Copyright 2017 RAND Corporation
**RAND**® is a registered trademark.

Cover: China's C919 airliner (promotional photo courtesy of Comac).

*Disclaimer: This research report was prepared at the request of the U.S.-China Economic and Security Review Commission to support its deliberations. Posting of the report to the Commission's website is intended to promote greater public understanding of the issues addressed by the Commission in its ongoing assessment of U.S.-China economic relations and their implications for U.S. security, as mandated by Public Law 106-398 and Public Law 113-291. However, it does not necessarily imply an endorsement by the Commission or any individual Commissioner of the views or conclusions expressed in this commissioned research report.*

### Support RAND
Make a tax-deductible charitable contribution at
www.rand.org/giving/contribute

www.rand.org

# Preface

The U.S. aerospace industry is a major contributor to U.S. exports and national security. China maintains industrial policies aimed at creating a globally competitive aviation industry. Building on previous studies by the RAND Corporation—*Ready for Takeoff: China's Advancing Aerospace Industry* and *The Effectiveness of China's Industrial Policies in Commercial Aviation*—the U.S.-China Economic and National Security Review Commission (USCC) asked RAND to assess Chinese investment in U.S. aviation. In this report, the term *aviation* generally refers to the industry of manufacturing aircraft and does not extend to operating airlines, which is not currently threatened by Chinese competition.

Given the economic and security sensitivities of Chinese investment in U.S. aviation, the USCC asked RAND to provide the context for China's future demand for aviation products and an update on China's aviation industrial policies and the state of its aviation industry; to review Chinese investment in U.S. aviation and related university connections with Chinese entities; and to assess the implications of the resulting technology transfer on U.S. national security and aviation industry competitiveness.

This research was conducted within the International Security and Defense Policy Center of the RAND Corporation's National Security Research Division (NSRD). NSRD conducts research and analysis on defense and national security topics for the U.S. and allied defense, foreign policy, homeland security, and intelligence communities and foundations and other nongovernmental organizations that support defense and national security analysis.

For more information on the International Security and Defense Policy Center, see www.rand.org/nsrd/ndri/centers/isdp or contact the director (contact information is provided on the web page).

# Contents

# Figures and Tables

## Figures

## Tables

# Summary

## Background

The U.S. aerospace and defense-manufacturing sector is a major contributor to the U.S. economy. It accounts for an estimated 13 percent of total U.S. manufacturing and, in 2015, it generated a $67 billion trade surplus for the United States. Roughly half of that is civil aviation, which is an open, globally competitive market. Boeing's large commercial aircraft (LCA) is a major component of civil aviation exports, but U.S. companies are also integral elements of global supply chains supporting regional jet (RJ) manufacturers Bombardier and Embraer. The United States is also a major manufacturer of and market for general aviation (GA) equipment such as business jets, small aircraft, and helicopters.

In March 2016, the People's Republic of China (PRC) issued its 13th Five-Year Plan reiterating support for the development of the aviation industry. The plan specifically mentioned LCA, RJ, and GA. In June 2016, Chengdu Airlines began the first commercial operations of an ARJ-21 RJ manufactured in China. The first test aircraft of the Chinese C919, a potential direct competitor to the Boeing 737, was rolled out in November 2015, but first flight is not anticipated until early 2017. China has an underdeveloped domestic GA market because of flight regulations and limited small airport infrastructure, but the Chinese State Council hopes to double its size by 2020.

## Findings

Since 2005, Chinese companies have steadily increased investment in U.S. aviation by acquiring, merging, or establishing joint ventures with more than a dozen U.S. aviation companies (see Figure S.1) without directly running afoul of U.S. regulation. Over the past decade, we identified from open sources on average one to two investments in U.S. aviation per year, including 12 mergers and acquisitions, three joint ventures, and nine other agreements or failed deals. The combination of Chinese government policy to become globally competitive in aviation and the availability of capital drives these investments, but they are constrained by U.S. government foreign investment and export laws as well as classic business concerns about return on investment.

**Figure S.1**
**Timeline of Chinese Investments in U.S. Aviation**

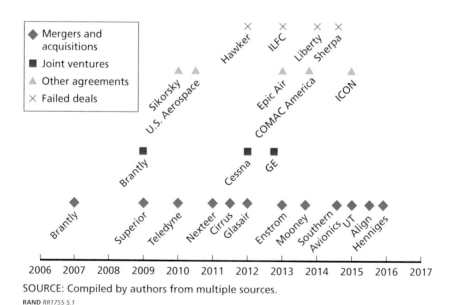

SOURCE: Compiled by authors from multiple sources.
RAND RR1755-S.1

For example, in 2011, a subsidiary of the Aviation Industry Corporation of China, a major Chinese state-owned enterprise (SOE) and defense contractor, wholly bought Cirrus Aircraft, the world's largest manufacturer of piston-powered GA aircraft. While none of the transactions appear to violate export controls or foreign investment regulations, they raise technology-transfer concerns that might have national security or competitiveness implications.

While China has unambiguous government policies supporting the development of a globally competitive aviation industry, Chinese investment in U.S. aviation over the past decade has primarily involved lower-technology GA manufacturers that do not affect U.S. competitiveness. Chinese demand for LCA may be as much as one-fifth of global demand, but the duopoly nature of global aviation also creates barriers to China's goal of developing a globally competitive commercial aircraft manufacturer, as any manufacturer of a new commercial aircraft struggles to achieve efficiencies of scale. While China continues to make more investments in U.S. aviation, it has not significantly expanded them beyond GA investments. While Commercial Aircraft Corporation of China (COMAC), a majority Chinese government-owned and government-controlled enterprise, is moving forward with the production of the ARJ21 and the development of the C919 with numerous U.S. partners, Chinese investors, including SOEs, have opportunistically acquired U.S. GA companies that fall outside export controls and U.S. foreign-investment regulations as GA technologies are broadly available.

Our main findings are:

- China will likely account for up to one-fifth of global demand for LCA and is trying to grow its domestic GA industry, which is currently underdeveloped.

- China has unambiguous policy driving a whole-of-government effort to develop a globally competitive aviation industry by producing LCA and expanding China's domestic GA market.

- Chinese investments in U.S. aviation have grown in scope and quantity over the past decade but are limited to smaller GA companies with technologies not particularly relevant to commercial or military aircraft, likely because of effective U.S. export and foreign investment regulations.

- There are few special relationships between Chinese institutions and U.S. universities related to aviation beyond the normal presence of Chinese graduate students attending U.S. aerospace programs and existence of university-wide study-abroad and cultural exchanges.

- Given the GA nature of most of the investments by Chinese firms to date, there are few technology-transfer concerns. The main benefits to China's industry would be on the business-process side, such as international marketing, achieving Federal Aviation Administration safety certifications, and product support.

- U.S. competitiveness is unlikely to be threatened in the near-term because production of China's LCA—the C919—may be further delayed and operate less efficiently than current Western narrow-body aircraft on the international market. However, some experts remain concerned about the transfer of engine or avionics technology through COMAC C919 joint ventures with Western companies; others think technology transfers are unlikely given U.S. export controls.

- A more competitive civil aviation industry broadly supports Chinese military aviation (e.g., larger talent pool, scales of efficiency, greater supply chain options). However, direct military implications are minimal because advanced commercial aviation technology differs from military aviation technologies (e.g., stealth, radar, supersonic engines).

# Acknowledgments

The authors would like to thank the U.S.-China Economic and Security Review Commission for its support for this research. We would also like to thank Seth G. Jones, director of the International Security and Defense Policy Center at RAND, for his advice and guidance; Scott Warren Harold, associate director, Center for Asia Pacific Policy at RAND; and Peder Andersen for their thoughtful and incisive reviews; and all of the aviation experts who shared their valuable time with us including Richard Aboulafia, Ronald Epstein, Sash Tusa, Joe Borich, and Richard Bitzinger, among others.

# Abbreviations

| | |
|---|---|
| AECC | Aero Engine Corporation of China |
| AIG | American International Group |
| AVIC | Aviation Industry Corporation of China |
| BUAA | Beijing University of Aeronautics and Astronautics |
| CAAC | Civil Aviation Administration of China |
| CAIGA | Chinese Aviation Industry General Aircraft |
| CALCE | Center for Advanced Life Cycle Engineering |
| CDB | China Development Bank |
| CEPREI | China Electronic Product Reliability and Environmental Test Research Institute |
| CFIUS | Committee on Foreign Investment in the United States |
| COMAC | Commercial Aircraft Corporation of China |
| CRHC | China Reform Holdings Corporation Limited |
| EASA | European Aviation Safety Agency |
| FAA | Federal Aviation Administration |
| FYP | Five-Year Plan |
| GA | general aviation |

| GDP | gross domestic product |
|---|---|
| GECAS | General Electric Capital Aviation Services |
| GIX | Global Innovation Exchange |
| ILFC | International Lease Finance Corporation |
| ITAR | U.S. Department of State International Traffic in Arms Regulation |
| JRI | Joint Research Institute |
| LCA | large commercial aircraft |
| MIIT | Ministry of Industry and Information Technology (People's Republic of China) |
| MOF | Ministry of Finance (People's Republic of China) |
| MOFCOM | Ministry of Commerce (People's Republic of China) |
| MOST | Ministry of Science and Technology (People's Republic of China) |
| MOU | memorandum of understanding |
| NASA | National Aeronautics and Space Administration |
| NCAMLTP | National Civil Aviation Medium- to Long-Term Plan (2013–2020) |
| NDRC | National Development and Reform Commission (People's Republic of China) |
| NUAA-JC | Nanjing University of Aeronautics and Astronautics-Jincheng College |
| PDB | Pudong Development Bank |
| PRC | People's Republic of China |
| R&D | research and development |
| RJ | regional jet |
| RMB | renminbi (China's currency) |

| | |
|---|---|
| SACS | Shanghai Aircraft Customer Service Company |
| SADRI | Shanghai Aircraft Design and Research Institute |
| SAIC | Shanghai Aviation Industrial (Group) Company |
| SAMC | Shanghai Aircraft Manufacturing Company |
| SASAC | State-Owned Assets Supervision and Administration Commission |
| SEI | Strategic Emerging Industries |
| SEIC | Shanghai Municipal Economic and Information Commission |
| SEZ | Special Economic Zone |
| SIG | Shanghai International Group |
| SOE | state-owned enterprise |
| UAC | United Aircraft Company |
| UCI | University of California, Irvine |
| USC | University of Southern California |
| USCC | U.S.-China Economic and Security Review Commission |
| USST | University of Shanghai for Science and Technology |
| Z-BEI | Shanghai Zhangjiang Berkeley Engineering Innovation Center |

# CHAPTER ONE

# Introduction

This report assesses five major issues:

- Chinese market demand for commercial and general aviation (GA) aircraft
- Chinese government policy and goals for its aviation industry
- Chinese investments in U.S. aviation companies and relationships with U.S. universities in the aviation arena
- possible technology transfer that would support Chinese aviation industry goals and military capabilities
- effect of all of the above on U.S. competitiveness.

## Sources and Methodology

We directly researched the first three topics using primary and secondary open-source materials on aviation markets, Chinese policy, Chinese investments in U.S. aviation companies, and university relationships. In Chapter Two, which covers aviation markets, we establish the significance of civil aviation to the U.S. economy, describe the global commercial aviation industry competitors by aircraft market and production rates, explore projections of Chinese demand for commercial aviation products, note the status of general aviation in China, and briefly examine the ability of China's industry to meet those needs. In Chapter Three, we review six major government policies that effect aviation and describe how the Commercial Aircraft Corporation of China's (COMAC's) C919 narrow-body aircraft development is being

funded and executed. In Chapter Four, we characterize Chinese investments in U.S. aviation or relationships with U.S. educational institutions over the past decade based on press releases and industry reports.
We catalogued all the ones publicly identified by major aviation-related
institutions—corporate and educational—in the United States and
China. In short:

- China will likely account for up to one-fifth of global demand for
  large commercial aircraft (LCA) and is trying to grow its domestic GA industry, which is currently underdeveloped.
- China has an unambiguous policy driving a whole-of-
  government effort to develop a globally competitive aviation
  industry by producing LCA and expanding China's domestic GA
  market.
- Chinese investments in U.S. aviation have grown in scope and
  quantity over the past decade but are limited to smaller GA companies with technologies not particularly relevant to commercial
  or military aircraft, likely because of effective U.S. export and
  foreign-investment regulations.
- There are few special relationships between Chinese institutions
  and U.S. universities related to aviation beyond the normal presence of Chinese graduate students attending U.S. aerospace programs and existence of university-wide study-abroad and cultural
  exchanges.

To address the possibility of technology transfers and the effect
on U.S. competitiveness, we spoke with nine subject-matter experts
who have knowledge about the aviation and aerospace industries, U.S.-
China business relations, U.S. trade restrictions and export controls,
and legal regulations on foreign investment in U.S. companies.[1] After
providing them with a summary of our findings on Chinese investments in U.S. aviation and Chinese government policy pertaining to
aviation, we solicited their insights on four broad aspects of Chinese

---

[1]   To maximize candor, we offered potential interviewees anonymity at their request. Our
guidelines were reviewed and approved by RAND's Human Subjects Protection Committee.

investments in U.S. aviation: (1) whether China had a strategy, (2) the potential for technology transfer accelerating the development of China's domestic aviation and aerospace industry, (3) the effect on U.S. global competitiveness, and (4) the implications for Chinese military capabilities. One of the authors of this report also attended the U.S.-China Aviation Summit in June 2016. In Chapter Five, we present those findings. In brief:

- Given the GA nature of most of the investments by Chinese firms to date, there are few technology-transfer concerns. The main benefits to China's industry would be on the business-process side such as international marketing, achieving Federal Aviation Administration (FAA) safety certifications, and product support.
- U.S. competitiveness is unlikely to be threatened in the near term because production of China's LCA—the COMAC C919—may be further delayed and operate less efficiently than current Western narrow-body aircraft on the international market. However, some experts remain concerned about the transfer of engine or avionics technology through COMAC C919 joint ventures with Western companies; others think technology transfers are unlikely given U.S. export controls.
- A more competitive civil aviation industry broadly supports Chinese military aviation (e.g., larger talent pool, scales of efficiency, greater supply chain options). However, direct military implications are minimal because advanced commercial aviation technology differs from military aviation technologies (e.g., stealth, radar, supersonic engines).

## Scope and Limitations

This report focuses on China's interaction with the U.S. aviation manufacturing industry. The aviation industry is global, and aviation is only one component of the broader U.S. aerospace industry. Even with effective U.S. export and foreign-investment controls, Chinese companies may find access to aviation technologies through other foreign non-U.S. companies. U.S. aviation export controls are well established with reasonably clear differentiation between commercial and military technologies. Space technology is far more dual use in nature, and U.S. space-technology export rules were changed significantly in 2015. Space-technology investments by China were beyond our scope.

Finally, we looked at legal or accidental technology transfer. We did not attempt to assess technology stolen through government or corporate espionage or cybercrime. Given the confidential nature of those issues to both companies and the government, it would be difficult to do a comprehensive open-source review of those issues.

# CHAPTER TWO

# Aviation Markets

Aviation industries are not only major economic drivers in their own right but also are status symbols of a developed economy and ultimately expressions of national power. For these reasons, China's government has long advocated for the development of the country's aviation industry and admired U.S. leadership in aviation. In this chapter, we outline the importance of the global aviation markets to United States, describe the nature of global commercial aviation manufacturing industry, discuss future market projections for the world and China, and briefly highlight aspects of the GA market in China. This background material provides necessary contextual details for understanding Chinese aviation policy and Chinese investments in U.S. aviation.

## United States Aerospace and Defense Industries

Fractions of U.S. production, manufacturing employment, trade exports in general, and trade exports with China are some indicators of the relative economic significance of U.S. aviation industries. The U.S. aerospace and defense industries, which include civil aircraft, military vehicles, and space systems, are global leaders and a significant component of the U.S. economy. It is estimated that the U.S. aerospace and defense industries contribute annually more than $300 billion in economic value—or 1.8 percent of gross domestic product (GDP).[1] Of

---

[1]  Brendan O'Neil, Shane Norton, Leslie Levesque, Charlie Dougherty, and Vardan Genan-yan, *Aerospace and Defense Economic Impact Analysis: Aerospace and Defense Economic*

that, roughly $90 billion is direct value added based almost completely on manufacturing of aerospace and defense products. The remainder comes from indirect value supporting those products—in other words, the supply chain and induced economic value because of the industries' existence. The direct value-added components of the aerospace and defense industries account for more than 1 million jobs, or 13 percent of the U.S. manufacturing industrial base. Roughly half of that economic value is civil and commercial systems, and the other half is defense and national security.

Total U.S. goods exports in 2015 were roughly $1.5 trillion, with aerospace exports accounting for 8 percent of all U.S. exports. In terms of aerospace imports and exports, the United States has a positive trade balance of $67 billion from exporting $126 billion of products in 2015.[2] The U.S. goods trade deficit was $745 billion.[3] Based on those numbers, aerospace and defense exports are close to one-tenth of U.S. exports, and the sector accounts for a trade surplus that is also roughly one-tenth of the current trade deficit.

In 2015, civilian aircraft accounted for 13 percent of total U.S. exports to China. In 2015, the United States exported $116 billion of products to China but imported $468 billion from China for an annual $345 billion deficit. Civilian aircraft, engines, and parts accounted for more than $15 billion in exports that year.[4] The civil aviation exports to China are up from just under $5 billion in 2006 and above the ten-year average of more than $8 billion. U.S. exports of noncommercial aircraft, engines, and parts, while growing, are a small share of civil aviation exports to China. In 2015, annual aviation imports from China to the United States amounted to less than $1 billion. For context, computers, electronics, appliances, and machinery dominated

---

*Impact Analysis: A Report for the Aerospace Industries Association*, IHS Economics, April 2016.

[2] Aerospace Industries Association, *U.S. Aerospace Trade Balance*, 2016.

[3] Bureau of Economic Analysis, Department of Commerce, "U.S. Trade in Goods and Services, 1992–Present," July 6, 2016.

[4] U.S. Department of Commerce, Census Bureau, "U.S. International Trade Statistics, 2016," 2016.

imports from China and accounted for more than $240 billion, or more than half of the $468 billion of U.S. imports from China. For clothing, footwear, furniture, linens and textiles, and sporting goods and toys, each category accounted for more than $15 billion of additional imports from China, accounting for much of the remainder of U.S. imports from China.

## Civil Aviation Markets

Civil aviation (i.e., nondefense aviation) is dominated by scheduled airlines flying LCA or regional jets (RJs). GA covers the rest of civil aviation, including unscheduled commercial, private, and government aviation such as business jets, light aircraft, and helicopters. Aviation is dominated by LCA, which account for more than 80 percent of commercial aircraft produced annually, as detailed in the following discussion. RJs are second, but they are produced at only one-tenth of the quantity of LCA. Although GA airframes may be more numerous in quantity than even LCA, their combined value is only on par with annual RJ revenues. While civil aviation is not just LCA, market success in the LCA category indicates aviation industrial strength and competitiveness.

LCA are typically defined as passenger jets that seat more than 100 passengers. Airbus and Boeing currently dominate the LCA market, splitting the market roughly in half. LCA are typically subdivided into narrow- and wide-body aircraft classes, also called *single-aisle* and *multi-aisle aircraft*. All of the narrow-body Boeing 737 variants and the Airbus 320 family of aircraft are produced in the greatest quantity of any commercial aircraft today. In 2015, Boeing delivered 495 variants of the 737, which seats 126 to 200 passengers.[5] Airbus delivered 491 of its Airbus 320 family, which seats 107 to 185 passengers.[6] In 2015, Boeing delivered an additional 267 wide-body aircraft

---

[5]   Boeing, Boeing commercial, homepage, undated.

[6]   Airbus, "Airbus Results 2015," website, 2016.

(e.g., 777, 787, 747 variants) for a total of 762, and Airbus delivered an additional 144 wide-body aircraft for a total of 635.[7]

The much smaller RJ market is primarily divided between Embraer of Brazil and Bombardier of Canada. In 2015, Embraer produced just more than 100 RJs, 80 percent of those seating 70–88 passengers and the rest seating 98–124 passengers.[8] In 2015, Bombardier delivered 44 RJs seating 66–104 passengers.[9] Both companies also build business jets and turboprop passenger aircraft. In the RJ market, there are a number of new competitors. Sukhoi of Russia produces the Superjet 100 (seating 87–108), of which roughly 125 have been produced cumulatively since 2008. Production of the Superjet 100 in 2015 appears to be less than half of the 37 aircraft produced in 2014.[10] Most of the Superjet 100s have been delivered to Russian airlines, but they are certified by the European Aviation Safety Agency (EASA) and operated by airlines in a number of countries other than Russia, such as Mexico and Ireland. At the end of 2015, COMAC delivered its first ARJ-21 (seating 78–90 passengers) to Chengdu Airlines, which began commercial operations in June 2016. Mitsubishi Aircraft Corporation's MRJ, which seats 70–90 passengers, saw first flight in 2015 with deliveries scheduled for late 2016. Finally, the Ukrainian company Antonov has produced an An-148 since 2009, but less than half of the 29 produced units are in commercial operation, most recently in North Korea, Russia, Cuba, and Ukraine.[11]

The duopolistic structure of global-aviation manufacturing— where two producers dominate each market sector—is not artificial, but it is the natural state of equilibrium for the industry in each product class for a number of reasons. First, annual production rates are relatively low compared with other industries such as automobiles or

---

[7]  Boeing, undated; and Airbus, 2016.

[8]  Embraer, "Embraer Releases Fourth Quarter and Fiscal Year 2015 Results and 2016 Outlook," São Paolo, March 3, 2016.

[9]  Bombardier, "Commercial Aircraft Status Reports," website, undated.

[10]  Russian Aviation Insider, "SSJ 100 Production Rates Are Down 54%," website, October 5, 2015.

[11]  Planespotters, "Antonov An-148 Operators," website, undated.

computers. Only the top one, two, or three producers can achieve competitive efficiencies of scale in production and global product support. Second, there are significant barriers to entry not only for competitors but also for new products in general. Development timelines for a new aircraft are on the order of a decade, and the financial commitment is in the billions of dollars—roughly on the order of $10 billion for a new LCA. While each country has its own aviation-safety authority, the FAA and EASA are the universally recognized leaders in aircraft certification; their certifications are widely accepted by agencies in other countries. However, the FAA and EASA certification processes are demanding. Not only do they require thorough flight testing to verify modeling and simulation, but they require adequate supply-chain quality assurance and component tracking such that future mishap investigations can differentiate between design flaws and quality control failures. Third, aircraft-operational efficiency and safety are critical to the airlines. Aircraft cost of ownership is less than 20 percent of the cost per seat for the airlines, which means that fuel, maintenance, and personnel make up the bulk of the costs.[12] Airlines that operate the third or fourth most-efficient aircraft have trouble remaining cost competitive in the long term. Given the three factors of scales of efficiency, barriers to entry, and a preference to operate the most-efficient aircraft, the duopoly structure of the industry is likely to continue.

In the early 1970s, Boeing's market share was more than 80 percent of the global (non-Soviet) aviation market. By the mid 1970s, Boeing only had 40 percent because of competition with McDonnell Douglas and Lockheed, who had 40 percent and 20 percent, respectively. By 1990, Boeing again peaked at 70 percent, with McDonnell Douglas and Airbus splitting the rest and Lockheed having left the commercial aircraft business.[13] Ultimately, McDonnell Douglass merged with Boeing, and Airbus now splits the market with Boeing, as detailed above. While a complete discussion of Airbus's success is

---

[12] Scott McCartney, "How Airlines Spend Your Airfare," *Wall Street Journal*, June 6, 2012.

[13] U.S. Congress, Office of Technology Assessment, *Competing Economies: America, Europe, and the Pacific Rim*, Washington, D.C.: U.S. Government Printing Office, OTA-ITE-498, October 1991.

beyond the scope of this research, it is worth noting that its success involved more than government intervention. Europe's aviation industrial base was a world leader in the first half of the 20th century. However, that industry was divided by numerous national borders, and the Soviet occupation of Eastern Europe further diminished the continental aircraft market, which was already smaller than the North American aircraft market. While Airbus's history does involve government intervention, its success was largely driven by consolidation of the European aviation industry along with the globalization of aviation markets.

Given global competition, it is highly likely that there will always be a second most-efficient manufacturer that can achieve some scales of efficiency. Historically, the leaders of the duopoly have changed over time with the introduction of innovative new products. Major aerospace companies such as the historical, now-merged Lockheed Corporation and McDonnell Douglas eventually left the commercial market when they found themselves on the losing side of those trends. Likewise, numerous RJ manufacturers have had aspirations of building larger aircraft in the narrow-body class of aircraft. Most-recent RJ models have just exceeded the 100-seat level but are still not as large as the Boeing 737s or Airbus 320s. Lacking technologies developed by Boeing and Airbus (e.g., advanced composites), these competitors are unlikely to produce an aircraft that is as cost-efficient per seat-mile. Given that challenge, RJ manufacturers struggle to make the business case for raising the capital necessary to challenge the LCA leaders.

A major motivation for this report and the continued observation of China's commercial aircraft–manufacturing efforts is China's potential ability to upset this duopoly equilibrium by operating on a non-commercial basis, providing substantial government support to its aviation industry, and absorbing losses that would dissuade new market entrants operating on a purely private-enterprise basis. China's unambiguous industrial policies and the scale of its aircraft market—both discussed in the next section—and its apparent political will to absorb huge losses to establish a presence in this industry give it a chance to succeed where the Japanese and Brazilians have not.

## Future Market Projections and China

Boeing and Airbus annually produce 20-year market outlooks for commercial aircraft. While the outlooks are understandably optimistic, they are projections of the market based on straight-line projections of current trends, which are generally correct until an exogenous shock such as 9/11 or the 2008 global financial crisis resets the trend line following a dip in commercial airline passenger traffic. In 2015, Boeing projected total global LCA deliveries over the next 20 years at 35,560.[14] Airbus projections are slightly less optimistic but similar, with an estimate of 32,585 LCA deliveries over the same time.[15] The Boeing projection means an average yearly global demand of 1,780 commercial aircraft, roughly 27 percent more than the combined current production annual rate of 1,397 in 2015. For comparison, the Boeing report also projects RJ deliveries at a total of only 2,490 over 20 years—or only 7 percent of the commercial aircraft market.

Boeing and Airbus see 39 percent of LCA demand over the next two decades coming from Asia. Boeing reports that half of that will be from China alone. China is projected to be purchasing an average of more than 300 commercial aircraft annually over the next 20 years, which would almost triple the size of the fleet in China to more than 7,200 aircraft from 2,570 in 2014. In a 2011 report, RAND did a detailed analysis of potential growth in China's commercial aircraft fleet.[16] Air-passenger traffic growth is driven by economic growth and transportation patterns, with air transportation competing with other modes of transportation such as rail, bus, or private automobile usage. Based on these circumstances, meeting current Boeing estimates would require maintaining the same transportation pattern, which is between higher U.S. and lower Japanese air-transport consumption levels, along with a 6-percent annual GDP growth. In that assessment, we suggest that Chinese population densities, continued urban-

---

[14] Boeing, "Current Market Outlook 2015–2034," website, 2015.

[15] Airbus, "Global Market Forecast: Flying by Numbers 2015–2034," website, 2016.

[16] Roger Cliff, Chad J. R. Ohlandt, and David Yang, *Ready for Takeoff: China's Advancing Aerospace Industry*, Santa Monica, Calif.: RAND Corporation, MG-1100-UCESRC, 2011.

ization, and investment in rail, particularly high-speed rail, will shift patterns toward using less air travel. With the recent slowdown in economic growth, some are skeptical that China will maintain its current 6-percent annual GDP growth through 2036.

Regardless, the optimistic case of industry projections suggest that China will be one-fifth of global demand for commercial aircraft; the less-optimistic case (slower economic growth or an increased preference for other transportation modes) would decrease projections to one-sixth. In either case, the numbers represent a significant fraction of global demand and is certainly an important market to Boeing. However, given the duopoly nature of the commercial aircraft market, one-fifth is not enough to achieve economies of scale in production. Even if COMAC captures 100 percent of the Chinese market, Boeing and Airbus could split the remaining four-fifths of the global market and still produce at twice the rate of COMAC. The prospect that COMAC could capture 100 percent of the Chinese domestic market is also unlikely, even with significant government protection. It would mean a monopoly within China, which would likely lead to significant cost growth in aircraft purchases and sustainment costs.

To ultimately upset the current aviation-industry equilibrium, COMAC would need to produce a competitive commercial aircraft product and market it successfully on a global scale. This means not just developing the technologies necessary to compete with Boeing and Airbus but also achieving safety certification from the FAA or EASA and establishing global supply chains to meet global-sustainment requirements. It would likely take China at least several decades to achieve this. In the meantime, if China proceeds in this direction, its aviation industrial policies will likely distort the global aviation market.[17]

---

[17]  For an example of aircraft price distortion is the Province of Quebec's investment in Bombardier and the potential distortion of RJ prices, see Jens Flottau, Michael Bruno, Graham Warwick, Guy Norris, and Bradley Perrett, "Subsidy Battle Anew," *Aviation Week and Space Technology*, July 4–17, 2016.

## General Aviation in China

While commercial aviation revenues are far more significant than GA, GA is often considered a potential growth market in China. GA, defined as unscheduled civil aviation, takes many forms, such as private pilots, executive business jets, air ambulances, forest firefighting, crop dusting, sky cranes, and tourism. In the United States, there is a long history of GA, which starts with the Wright brothers in 1903. Moreover, GA holds a prominent place in U.S. society: In many states, Americans can get their pilot's license at a younger age than they can their driver's license. Additionally, an extensive infrastructure of thousands of local airports support GA.[18] In China, by contrast, the military controls all the airspace, and there is neither public familiarity with GA nor substantial infrastructure to support it. In May 2016, the Chinese State Council released guidelines on furthering the development of GA in China.[19] The guidelines set 2020 goals of doubling the current number of GA airports with an additional 300, relaxing the altitude restrictions that require a flight plan, and accelerating flight-plan approval from weeks/days to possibly hours—all in line with international GA norms.[20] It may enable the expansion of GA for the wealthy and corporate entities in China, but it is unlikely to change the cultural bias of government control or affect a large fraction of the population.

## Civil Aviation Manufacturing in China

Civil aviation manufacturing is largely the same as represented in the 2011 RAND report *Ready for Takeoff: China's Advancing Aerospace Industry.*

---

[18] Federal Aviation Administration, "General Aviation Airports: A National Asset," website, May 2012.

[19] The State Council of the People's Republic of China, "China to Boost General Aviation," May 17, 2016.

[20] James Fallows, *China Airborne*, New York, N.Y.: Pantheon Books, 2012.

Except possibly in the case of helicopters, China's current ability to meet demand with indigenous aircraft is limited. Its indigenous regional jet, the ARJ21, will begin deliveries in 2011, but the regional jet market in China is small. China's indigenous large commercial aircraft, the C919, will not begin deliveries until the middle of the decade, at the earliest, and it will be a narrow-body aircraft that competes only with the Boeing 737 series and Airbus A320 series. All wide-body aircraft will be imported at least through 2020.[21]

In 2016, China has made progress on all these accounts and continues to push them forward, but it has not met its planned schedule. ARJ21 deliveries only began in late 2015, and commercial operations started in June 2016. The first C919 was rolled out in October 2015, but it is unclear whether first flight will occur in 2016, with deliveries to start at the earliest a year later.

More significantly, as of June 2016, the ARJ21 has yet to be certified by the FAA, and FAA certification of the C919 is even less advanced, as the C919 has not seen first flight yet. To date, COMAC has not pursued EASA certification, as FAA certification would be broadly accepted, including by European authorities. Recently, the latest version of the Harbin Y-12 was FAA certified, but it is a twin-engine turboprop utility aircraft based on a Western aircraft design that went out of production in 1988.[22] In 2015, COMAC began preliminary design work with Russia's United Aircraft Company (UAC) on wide-body aircraft designs.[23]

Table 2.1 and Figure 2.1 show the suppliers and joint ventures associated with the C919, which help to assess the state of the art in the Chinese domestic civil aviation sector. While the C919 structure, fuselage, wings, and tail are produced in China, Western partners provide the C919 engines, avionics, control systems, communications, land-

---

[21] Cliff, Ohlandt, and Yang, 2011, p. xii.

[22] Lena Ge, "China-Made Y-12F Turboprop Aircraft Gets FAA Type Certification," *China Aviation Daily*, February 25, 2016.

[23] Maxim Pyadushkin and Bradley Perrett, "Russo-Chinese Widebody Concept Design Underway," *Aviation Week and Space Technology*, February 11, 2015.

**Table 2.1**
**Selected International Partners in the C919 Program**

| C919 Program Partners | Contribution |
|---|---|
| U.S. partners | |
| Alcoa | Fasteners, metal castings, and components |
| Eaton Corporation | Pipelines for fuel and hydraulic systems |
| General Electric | Engines (CFM International); engine nacelle, thrust reversers (Nexcelle); avionics system core processing and display; onboard maintenance and flight-data recording |
| Goodrich Corporation | Exterior lighting; landing gear and engine nacelle components |
| Hamilton Sundstrand | Electric-power generation and distribution; cockpit pilot controls (e.g., side sticks, pedals) |
| Honeywell International | Flight-control system; auxiliary power unit; wheels and tires, braking system; inertial reference and air-data systems |
| Kidde Aerospace (Hamilton Sundstrand subsidiary) | Fire and overheat protection systems |
| Parker Aerospace | Fuel and hydraulic systems (NEIAS Parker); flight-control actuation (Parker FACRI) |
| Rockwell Collins | Communication and navigation systems; integrated surveillance system; cabin core system |
| Non-U.S. partners | |
| Fisher Advanced Composite Components (Austria) | Cockpit, cabin interior, kitchens, restrooms |
| Liebherr Aerospace Toulouse (France) | Air-management system |
| Safran (France) | Engines (CFM International), in-flight entertainment |
| Thales (France) | Electrical wiring interconnection system (Shanghai SAIFEI); engine nacelle, thrust reversers (Nexcelle); propulsion (CFM International) |
| Liebherr Aerospace Lindenberg (Germany) | Undercarriage system; landing-gear system (Liebherr LAMC) |

SOURCE: Based on Cliff, Ohlandt, and Yang, 2011, and updated.
NOTE: See Airframer, "COMAC C919," website, May 5, 2016, for an extensive list.

**Figure 2.1**
**Infographic of C919 Partners**

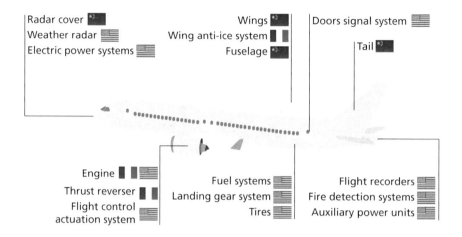

Radar cover
Weather radar
Electric power systems

Wings
Wing anti-ice system
Fuselage

Doors signal system
Tail

Engine
Thrust reverser
Flight control
actuation system

Fuel systems
Landing gear system
Tires

Flight recorders
Fire detection systems
Auxiliary power units

SOURCE: Illustration re-created based on figure in Steven Jiang, "China to Take on Boeing, Airbus with Homegrown C919 Passenger Jet," CNN.com, November 2, 2015.
RAND RR1755-2.1

ing gear, and more. As China's SOEs produce all those components for indigenous military aircraft, this suggests that China's aviation-manufacturing industry is not competitive in these areas.

In August 2016, the engine unit of the Aviation Industry Corporation of China (AVIC) split off into a new SOE, the Aero-Engine Corporation of China. The entity allows for more direct investment and control by the central government and Beijing regional authorities, most likely with the goal of developing a more efficient and competitive engine supplier to avoid continuing foreign dependence. As AVIC emphasizes military aircraft, the spin-off may also suggest an increased focus on commercial jet engines to support COMAC in the future.[24]

---

[24] Bradley Perrett, "Aero Engine Corp. of China Inaugurated, Separated from Avic," *Aviation Week and Space Technology*, September 2, 2016b.

## Findings on Aviation Markets

- Civil aviation is a significant component of the U.S. economy and U.S. exports.
- LCA dominate the civil aviation markets over RJs and GA.
- Both LCA and RJ markets tend toward duopolies because of scales of efficiency in production and global product support and barriers to new product entry.
- China will likely account for up to one-fifth of global demand for LCA and is trying to grow its domestic GA industry, which is currently very underdeveloped.
- Chinese global competitiveness in civil aviation is limited based on the C919's dependence on foreign suppliers.

The next chapter explores Chinese government policy toward its domestic aviation industry and provides some examples of how it is implemented.

# CHAPTER THREE

# China's Government Policy for Commercial Aviation

China adopts a whole-of-government approach to promoting its domestic commercial aviation industry. Such an approach begins with broad declarations of policy priorities from the Chinese State Council, the substance of which is fleshed out in the form of policy-planning documents and directives by the various ministries that seek to accomplish specific Chinese development goals. Local government entities subsequently work with Chinese state-owned enterprises (SOEs) and state development banks to fashion preferential industrial policies to encourage indigenous innovation and technological breakthroughs in commercial aviation development.

This chapter examines Chinese government policy for commercial aviation development and highlights the plans and documents released by the Chinese government related to the promotion of commercial aviation. The chapter continues with an analysis of the processes of policy implementation at the national- and local-government level, using the example of the COMAC C919 narrow-body jetliner currently under production as a case study.

While China has unambiguous government policies working toward making its aviation industry globally competitive, most of those efforts focus on LCA manufacturing and COMAC's C919. In the next chapter, Chapter Four, we will see that most Chinese investments in U.S. aviation are not particularly relevant to LCA and unlikely to affect U.S. global competitiveness.

## Chinese Government Planning and Commercial Aviation Development

There are six key government plans relevant to commercial aviation development in China:

1. Five-Year Plans (FYPs)
2. Strategic Emerging Industries (SEIs) documents
3. National Civil Aviation Medium- to Long-Term Plan (2013–2020) (NCAMLTP)
4. National Medium- and Long-Term Program for Science and Technology Development (2012–2030)
5. 12th Five-Year Plan for Chinese Civil Aviation Development
6. Made in China 2025 initiative.

### Five-Year Plans

FYPs are produced by the Chinese State Council and represent development priorities across a range of industries within China. The commercial-aviation manufacturing industry, sometimes referred to as "high-technology transportation equipment," has been featured as a development priority since the tenth FYP in 2001.[1] The 13th FYP, released in March 2016, mentions "aerospace equipment" as the first of seven "advanced equipment innovation development projects," which calls for "breakthroughs in civilian aviation engine technology" and the "acceleration of research in large-body aircraft, helicopters, regional jets and general aviation."[2] FYPs do not provide specific directives or details on how China seeks to execute such goals. Rather, they provide a broad outline of priorities for China's various ministries to implement.

---

[1]  People's Republic of China, National People's Congress, *China's 10th (2001–2005) Five-Year Plan*, Beijing, 2001; People's Republic of China, National People's Congress, *China's 11th (2006–2010) Five-Year Plan*, Beijing, 2006; People's Republic of China, National People's Congress, *China's 12th (2011–2015) Five-Year Plan*, Beijing, March 2011; and People's Republic of China, National People's Congress, *China's 13th (2015–2020) Five-Year Plan*, Beijing, 2015.

[2]  People's Republic of China, National People's Congress, "Work Report on 13th Five-Year Plan for the Country's Economic and Social Development [中华人民共和国国民经济和社会发展第十三个五年规划纲要]," March 17, 2016.

## Strategic Emerging Industries

In 2010 and 2012, the Chinese State Council released two key documents on SEIs. The first, called *The Decision on Accelerating the Cultivation and Development of Strategic Emerging Industries* (国务院关于加快培育和发展战略性新兴产业的决定), was released in October 2010 and laid out in broad strokes the strategy and rationale behind seven SEIs, including advanced equipment manufacturing.[3] The content of this document does not touch on specific development policies but rather presents the goals and aspirations of SEI policy in general.

The second document is called the *Development Plan for Strategic Emerging Industries of the 12th Five-Year Plan* (十二五国家战略性新兴产业发展规划). This document lays out in finer detail the development goals of the seven industries that Chinese policymakers have identified as drivers of Chinese innovation and development through 2020. In the document, "advanced equipment manufacturing" (高端装备制造产业) was identified as one of seven SEIs under which civilian aviation was highlighted as a key development goal. In particular, the document lays out specific goals for civilian aviation development for both 2015 and 2020, including the development of the C919, a single-aisle, 150-seat large-capacity jet as well as the development of the ARJ21 RJ.[4] Because of the level of specificity and origin from China's top policymaking body—the State Council of the People's Republic of China (PRC)—the 2012 SEI document represents arguably the most important Chinese government plan for China's civilian-aviation industry.

Two chapters of the 2012 SEI development plan merit closer examination for their explicit emphasis on domestic-aviation development. In section four of chapter three, titled "High-End Equipment

---

[3]   The seven SEIs are bioindustry (生物产业); new energy (新能源产业); advanced equipment manufacturing (高端装备制造产业); new materials (新材料产业); energy conservation and environmental protection (节能环保产业); new energy vehicles (新能源汽车产业); and next-generation information technology (新一代信息技术产业).

[4]   State Council of the People's Republic of China, "PRC State Council Notice on Program for the Development of Strategic Emerging Industries During the 12th Five-Year Program Period [国务院关于印发 '十二五' 国家战略性新兴产业发展规划的通知]," July 9, 2012.

Manufacturing," policymakers list "aviation industry" as the first priority of development:

> We need to devise a comprehensive scheme for conducting R&D [research and development] of aviation technology and developing and industrializing aviation products, exploring markets and providing services. We need to speed up the development of large passenger planes capable of competing in markets, promote the industrialized development of a series of advanced planes for regional airlines, and develop new-model regional planes; develop new-model planes and helicopters for general use that meet market needs, and build a general aviation industry. We need to achieve breakthroughs in developing core technologies pivotal to plane engines and expedite the process of industrializing the production of plane engines; expedite the development of services for maintaining aviation equipment and systems; and improve the aviation industry's core competitive capabilities and ability to achieve specialized development.[5]

In section 13 of chapter four, titled "Major Engineering Projects," the Chinese State Council notice is specific about SEI program goals for PRC aviation:

> In light of the requirement for safety, economy, comfort and environmental protection, we need to develop single-aisle, 150-seat C919 aircraft for main air routes. We need to speed up the process of brainstorming scientific-technological projects and develop the designs and technologies for building reliable, low-cost and digital aircraft for regional airlines and aircraft (including helicopters) for general use. We should push forward the scaled production and development of the ARJ21 aircraft series for regional airlines, support the remodeling of the Xinzhou aircraft series for regional airlines, develop new-type aircraft for regional airlines, and develop large jet aircraft for public services and new aircraft (including helicopters) for general use; expand the use of regional airports, and firmly push forward the pilot projects of launching

---

5   State Council of the People's Republic of China, 2012.

commuter air services. By the year 2015, China's capabilities of developing aviation equipment should have grown significantly.[6]

When the 2010 SEI decision document was released, the Chinese government set a target for the seven SEIs to increase their contribution to China's GDP from 2 percent in 2010 to 8 percent by 2015 and 15 percent by 2020.[7] The Chinese government reportedly allotted 4 trillion renminbi (RMB) (about $635 billion) in state funds to support the seven SEIs during the 12th FYP period (2011–2015).[8] Given the broad scope of the seven industries (e.g., high-end equipment manufacturing), and the debatable quality and availability of economic data in China, it is unclear what fraction of the SEI resources was devoted to aviation. It is also difficult to assess whether China achieved the target development goals for 2015. We do know that significant state resources were committed and that aviation technology—including the C919 and ARJ21 projects—was specifically identified.

## National Civil Aviation Medium- to Long-Term Plan (2013–2020)

Another important strategic guiding document for Chinese civilian-aviation development is the *National Civil Aviation Medium- to Long-Term Plan (2013–2020)* (民用航空工业中长期发展规划 [2013–2020 年]), which represents the only long-term plan that specifically targets the domestic aviation industry in China. Issued in 2013 by the Ministry of Industry and Information Technology (MIIT), the NCAMLTP calls for domestically manufactured civilian aircraft to account for no less than 5 percent of China's total civilian aviation market share by 2020 and sets a revenue target for China's domestic aviation industry of 100 billion RMB (about $15 billion) by the same year.[9] Given

---

[6]   State Council of the People's Republic of China, 2012.

[7]   "Strategic Emerging Industries: A New Economic Engine for Chinese Economic Development [战略性新兴产业: 中国经济发展的新引擎]," *China Economic Herald*, July 28, 2011.

[8]   "Strategic Emerging Industries: A New Economic Engine for Chinese Economic Development [战略性新兴产业: 中国经济发展的新引擎]," 2011.

[9]   "Middle- and Long-Term Development Plan for the Civil Aviation Industry (2013–2020) [国家中长期科学和技术发展规划纲要 (2012-2030年)]," Ministry of Industry and Infor-

ARJ21 and C919 delays, it seems unlikely that China will meet these targets. At a cost of roughly $100 million per aircraft for a standard 737 or A320 narrow-body aircraft, according to current Boeing and Airbus list prices, which are higher than confidential contract prices, COMAC would need to deliver 150 aircraft to meet its target revenue, which seems unlikely. At a more realistic $50 million per C919, the company would need to produce and sell 300 aircraft by 2020 to reach the target $15 billion in revenue. Such regional jets as the ARJ21 are roughly half the cost of an LCA, so two ARJ21s would produce roughly the same revenue as a single C919.

The NCAMLTP emphasizes the development of a domestic turbine-engine sector for narrow-body aircraft. For example, it calls for China to "accelerate the comprehensive establishment of an R&D center" and a "production and manufacturing line for jet engines by 2015." [10] The NCAMLTP sets the goal for China to become a technological innovator in RJ development—with the ARJ21 as the flagship platform—as well as an innovator in GA, multipurpose aircraft, and helicopters for domestic and international markets. Finally, the NCAMLTP offers a detailed blueprint for creating a system of aviation production, R&D and "innovation hubs" across different regions and cities in China.[11]

Because of the specificity of detail and because it was issued by the MIIT—a key government ministry in charge of China's civilian-aviation industry—the NCAMLTP should be considered, alongside the 2012 SEI document, as one of the most important Chinese government guidelines for China's commercial-aviation industry. MIIT's role in implementation of aviation policy is further discussed in the implementation of Chinese Government Aviation Policies section below.

---

mation Technology, May 22, 2013.

[10] "Middle- and Long-Term Development Plan for the Civil Aviation Industry (2013–2020) [国家中长期科学和技术发展规划纲要（2012–2030年)]," 2013.

[11] "Middle- and Long-Term Development Plan for the Civil Aviation Industry (2013–2020) [国家中长期科学和技术发展规划纲要（2012–2030年)]," 2013.

## National Medium- and Long-Term Program for Science and Technology Development (2012–2030)

Issued in 2012 by the Chinese State Council and Ministry of Science and Technology (MOST), the *National Medium- and Long-Term Program for Science and Technology Development (2012-2030)* (国家中长期科学和技术发展规划纲要 [2012–2030年]) offers broad priorities for the development of the Chinese commercial aviation industry. For example, under the category "Efficient Transport Technology and Equipment," the program calls for China to build a multipurpose general aviation aircraft. The program does not provide specifics for implementation or growth targets for the industry, however, and is regarded as a guiding document for broader science and technology priorities in China.

## 12th Five-Year Plan for Chinese Civil Aviation Development (2011–2015)

Finally, the *12th Five-Year Plan for Chinese Civil Aviation Development (2011–2015)* (中国民用航空发展第十二个五年规划 [2011–2015年]) is a policy-guideline document issued by the Civil Aviation Administration of China (CAAC) after the unveiling of the Chinese State Council's 12th FYP in March 2011. The plan calls for the implementation of strategy to enhance the "safety, popularity, and globalization of China's civilian aviation industry," as well as to "scientifically grasp development rules, actively adapt to environmental changes, effectively resolve various conflicts, and work harder to contribute more to China's civil aviation development."[12] Similar to the FAA, CAAC is China's largest government agency in charge of regulating air-traffic control, air safety, and air-transportation services for commercial aircraft. The CAAC drafts a variety of policies, development strategies, laws, rules, and regulations for civil aviation, including safety standards, personnel qualifications, industry operations, and certifications of civil aircraft pilots and operators. The CAAC also leads aviation negotiations and contracts with overseas partners.

---

[12] "12th Five-Year Plan for Chinese Civil Aviation Development [中国民用航空发展第十二个五年规划（2011-2015年)]," Civil Aviation Administration of China, April 2, 2011.

Arguably the most important function of the CAAC is its ability to approve all purchases of aircraft by Chinese airlines. This role gives CAAC significant influence over China's commercial airline industry by encouraging purchase of domestically produced jetliners such as the C919. It is also known, based on COMAC's news reports on its website, that the director of CAAC has visited COMAC's headquarters on several occasions, including in May 2016, when he expressed CAAC's "full support" for the manufacturing and production of the C919 aircraft.[13]

## Made in China 2025 Initiative

Finally, there is the Made in China 2025 initiative.[14] Released by the State Council on May 8, 2015, the document serves as the first ten-year program for China's manufacturing industry, with a focus on new technologies and innovation. The ultimate goal of this program is to transform China into one of the world's great powers in manufacturing by the 100th anniversary of the founding of the PRC in 2021. Among the program's priorities are aviation and aerospace technologies, which is evidence of continuing support from the State Council. However, Made in China 2025 provides few details, stating only that China should "speed up research and development of large aircraft, research and development of wide-body passenger aircraft, encourage international cooperation on researching and developing heavy helicopters." The document does not set any aviation-specific targets beyond the broad goals mentioned. While light on details, the Made in China 2025 program does not indicate continued government support for civilian aviation development through enhanced production and manufacturing practices.

China subsequently published a "Made in China 2025 Key Technology Roadmap" ("中国制造2025"重点领域技术路线图) in

[13] "CAAC Minister Feng Zheng Lin Visits COMAC Lab: Fully Supports C919 Research [中国民航局局长冯正霖到中国商飞调研：全力支持C919研制工作]," *COMAC News Center*, May 10, 2016.

[14] State Council of the People's Republic of China, "State Council Notice on Printing Made in China 2025 [国务院关于印发《中国制造2025》的通知]," May 8, 2015.

October 2015; released by MIIT, it set domestic production targets to include a section on aerospace. Some of the targets set forth in this document appear to be pulled from the targets in the National Civil Aviation Medium- to Long-Term Plan (2013–2020).[15] However, the Made in China 2025 Key Technology Roadmap also sets new targets for domestic aircraft, aircraft engines, and equipment. For example, the document says that, by 2025,

> annual revenue of domestic civil airlines should exceed 200 billion RMB; domestic trunk line airplane delivery should capture over 10 percent of the domestic market; regional turbo-prop delivery should capture 10 to 20 percent of the global market; and general aviation planes and helicopter delivery should capture 40 percent and 15 percent of the global market respectively.[16]

The document also calls for the CJ-1000A turbofan jet engine model to be completed by 2020 and "prepared for commercial use" by 2025.[17] Finally, the document calls for domestically produced regional aircraft components and general-purpose aircraft components to "capture 30 percent and 50 percent of the market share in 2025, respectively."[18]

---

[15] For example, the Made in China 2025 Key Technology Roadmap document calls for domestically manufactured civilian aircraft to account for no less than 5 percent of China's total civilian aviation market share by 2020 and sets a revenue target for China's domestic aviation industry of 100 billion RMB by the same year, among other targets. The same targets appear in the National Civil Aviation Medium- to Long- Term Plan (2013–2020) document.

[16] U.S.-China Business Council, "Unofficial USCBC Chart on Localization Targets by Sector Set in the MIIT Made in China 2025 Key Technology Roadmap," February 2, 2016, p. 3.

[17] U.S.-China Business Council, 2016, p. 3.

[18] U.S.-China Business Council, 2016, p. 4.

**Figure 3.1**
**Chinese Government Policy Coordination Process Supporting Commercial Aviation Development**

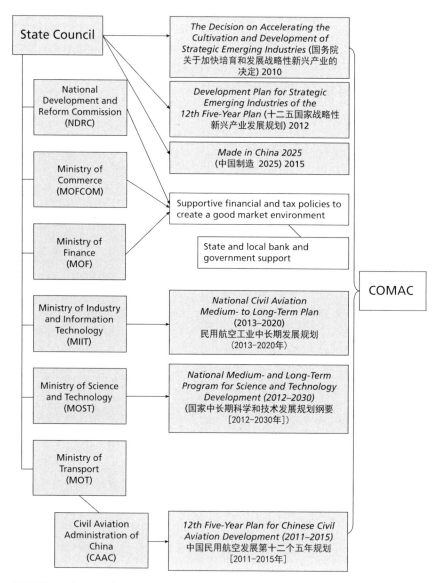

SOURCE: Author rendering based on open sources.
RAND RR1755-3.1

## Implementation of Chinese Government Commercial Aviation Policies

One of the most important functions that sets Chinese government policies apart from policies in other developed countries is their ability to direct China's ministries to pool resources and offer financial incentives, including direct investments, to SOEs and private industry to make technological breakthroughs and create a favorable market environment for commercial aviation innovation. Figure 3.1 shows how such a process might work, starting with the 2012 SEI policy document and with COMAC as its target recipient. State Council top-level policies drive the development of ministry-level policies. Those policies are coordinated to create market opportunities for and to direct funding resources from both government and banking to state-owned enterprises for execution (COMAC, in the case of aviation).

After the State Council's SEI document was released in 2012, the National Development and Reform Commission (NDRC)—a macroeconomic-management agency under the State Council—was chosen as the interministerial coordination group to implement SEI policies related to civilian-aviation development. The NDRC coordinates with the Ministry of Commerce (MOFCOM), MIIT, MOST, and the Ministry of Finance (MOF) to offer policy proposals that spur development and innovation. These ministries direct Chinese state development banks to offer a mix of preferential financial loans, tax incentives, investments in infrastructure, personnel training, as well as provide favorable policies that allow domestic aviation companies to enter into joint ventures with foreign partners. These ministries also work with local governments to create talent-incubator centers—such as R&D hubs and research labs within universities—to develop the domestic aviation industry.

The MIIT, through its NCAMLTP, also plays a particularly important role in directing resources and funding to developing target areas and industries based on the specifics of the NCAMLTP as well as its targeted goals for domestic commercial aviation production, revenue, and profit. As the ministry in charge of China's only long-term plan that specifically addresses China's aviation industry, the MIIT

**Figure 3.2**
**COMAC Shareholders and Subsidiaries**

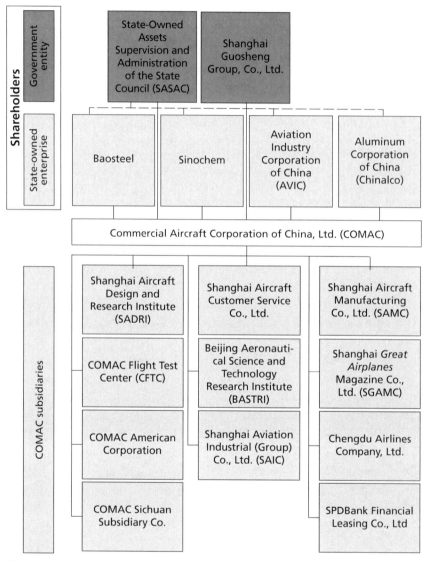

SOURCE: Generated from information on COMAC website.

wields unique influence over the direction of China's overall aviation development.

Finally, the main recipient of these government programs and processes is COMAC and, in particular, the C919 narrow-body jetliner currently in development (discussed in the next section). The C919 project provides a useful case study in understanding the crucial role that Chinese government policies serve in promoting China's domestic aviation industry.

## COMAC and the C919 Jetliner

In 2008, the Chinese government merged two of China's largest aviation manufacturers into Aviation Industry Corporation of China (AVIC) and spun off a state-owned commercial aircraft manufacturer, Shanghai-based COMAC. The merger was an attempt to position China as a major player in the global commercial aviation market.[19] COMAC is not publicly listed on any stock market. It is essentially a joint venture between the shareholders listed in Figure 3.2. Some of those shareholders are publicly listed or have components listed on stock markets including Baosteel, Sinochem, AVIC, and Chinalco. Others, such as State-Owned Assets Supervision and Administration Commission (SASAC) or the Guosheng Group, are special government entities that supervise government holdings of SOEs. In most cases, for large SOEs, the government holds a controlling interest in the company. As of August 2016, SASAC lists all four of the SOE shareholders of COMAC as SOEs it supervises.[20] Because SASAC is controlled by the State Council and the Guosheng Group is controlled by the Shanghai Municipal People's Government, COMAC is essentially a joint venture between China's central government and the Shanghai local government. Such arrangements are common in the Chinese system.

---

[19] "About Us [关于我们]," COMAC website, undated.

[20] State-Owned Assets Supervision and Administration Commission (SASAC), "List of Central Enterprises," SASAC website, August 3, 2016.

The C919 is COMAC's flagship narrow-body twin-engine commercial jetliner and was conceived to eventually compete with the Boeing 737 and Airbus A320, among other single-aisle airliners. Although COMAC had designs on selling the C919 internationally, Chinese carriers are virtually the only customers for the C919. As of May 2016, Chinese domestic airlines accounted for 334 of a total 344 orders, or 97 percent.[21] General Electric Capital Aviation Services (GECAS) is, to date, the only foreign commercial aircraft leasing company to have purchased the aircraft, ordering ten C919 aircraft in 2010.[22] Given that GECAS globally owns more than 2,000 aircraft, such a small order will probably only serve leasing customers within China.[23] The lack of purchases from foreign airliners and leasing companies is most likely the result of insufficient international accreditations. As of August 2016, the C919 has yet to receive FAA certification that COMAC has sought,[24] which, in turn, contributes to the lack of an operational track record or evidence of global product support. Without FAA or EASA certification and global-scale operations, the C919 is unlikely to be cost-competitive. Even with certification and global sales, the C919 will have to demonstrate that operational costs do not exceed any upfront savings in price. As previously mentioned, aircraft cost is less than 20 percent of the cost per seat for the airlines.

The process of investing in, manufacturing, and exporting a domestically built jetliner from scratch to compete with Boeing and Airbus is no small task and requires significant upfront capital. To address this challenge, the Chinese government has used local government agencies, state-backed banks, SOEs, and preferential tax and fiscal policies to achieve its development objectives. The following

---

[21] ABCDlist, "COMAC C919 Production List," ABCDlist website, June 7, 2016. Including letters of intent (LOIs) and option packages, a total of 521 C919s have been ordered.

[22] "China Unveils Jetliner in Bid to Compete with Boeing, Airbus," Associated Press, November 2, 2015.

[23] "China Wins 100 C919 Orders, Breaks Airbus-Boeing Grip," *Bloomberg News*, November 16, 2010.

[24] Siva Govindasamy and Matthew Miller, "Exclusive: China-Made Regional Jet Set for Delivery, but No U.S. Certification," Reuters, October 21, 2015.

sections highlight some of these actors and tools as they relate to the investment and manufacturing of the C919 aircraft.

## C919 Start-Up Capital

After the C919 project was approved by the NDRC in 2009, COMAC leveraged capital from Chinese SOEs and state development banks. This included 6 billion RMB ($876 million) from SASAC, 5 billion RMB ($730 million) from the Shanghai municipal government's Guosheng Group, and 1 billion RMB ($146 million) each from Aluminum Corporation of China (Chinalco), Baosteel Group, and Sinochem.[25] China's Bank of Communications provided a 30 billion RMB ($4.38 billion) line of credit. COMAC also relied on loans from China Development Bank (CDB) for seed funding.[26]

SASAC stands out as a key source of investment funds for the C919 project. In addition to the start-up funding, SASAC remains an ongoing source of government support for the C919. Based on news reports on COMAC's website, for example, SASAC officials frequently visit COMAC's Shanghai headquarters. Many of these officials come from provincial SASAC offices. In June 2016, for example, the director of Sichuan SASAC, Xi Jin, visited COMAC and toured the C919 manufacturing facilities. During his visit, COMAC officials thanked the Sichuan SASAC and government for their "continued support in jointly promoting the development of China's large commercial aircraft."[27] These and other meetings suggest that national and local government SASAC entities likely remain a key source of Chinese government support for the C919 beyond the initial 6 billion RMB

---

[25] Keith Crane, Jill E. Luoto, Scott Warren Harold, David Yang, Samuel K. Berkowitz, and Xiao Wang, *The Effectiveness of China's Industrial Policies in Commercial Aviation Manufacturing*, Santa Monica, Calif.: RAND Corporation, RR-245, 2014, pp. 25–26.

[26] Crane et al., 2014, pp. 25–26.

[27] Because of scarcity of data in some cases on specific levels of government support for the C919 project, we highlight COMAC site visits by Chinese government officials as important secondary sources on the degree of importance that China's government, development banks, and asset management companies place on the development of C919. See "Director of Sichuan SASAC Xu Jin Visits COMAC [四川省国资委主任徐进到中国商飞访]," *COMAC News Center*, June 6, 2016.

in upfront capital. For example, in 2015, the Export-Import Bank of China, which is "solely government owned and under the direct leadership of the State Council," signed a financing agreement with COMAC.[28]

The second key source of investment for the C919 project came from the Shanghai municipal government. In addition to the 5 billion RMB from the Shanghai municipal government's Guosheng Group, the Shanghai government successfully lobbied to have the manufacturing and R&D headquarters of the C919 located in the Zhangjiang Hi-Tech Park in Shanghai's Pudong district—one of China's most technologically innovative and wealthy districts.[29] The move of the headquarters to Pudong prompted the Shanghai municipal government to adopt a whole-of-government approach to promote the development of the C919 that leveraged local ministries, SOEs, and state-development banks.

COMAC also has been eager to attract foreign purchasers and investors. Between 2015 and 2016, senior COMAC officials met with officials from the United States, Russia, Iran, Singapore, Indonesia, Nigeria, and Israel, among other countries.[30] In June 2016, Russia and China signed an intergovernmental agreement to jointly develop a wide-body LCA through the United Aircraft Corporation of Russia and COMAC.[31] Sales of LCA to airlines are significant enough to often require the tacit approval of the airline's local government. If the country still operates national airlines, the government usually retains some form of direct veto over any purchase agreement. Even if the

---

[28] China Exim Bank, "Export-Import Bank of China," website, undated; and "COMAC Signs Financing Framework Agreement with TEIBC," COMAC news website, October 10, 2015.

[29] "China: Homegrown C919 Jets' Final Assembly Line Settles in Shanghai's Pudong," Xinhua News, November 19, 2009. Zhangjiang Hi-Tech Park is a Special Economic Zone (SEZ) and is known as China's "Silicon Valley," with more than 3,000 R&D institutions and more than 10,000 registered companies. See Zhangjiang Hi-Tech Park, "About Us," website, undated.

[30] COMAC website, "News Center," undated.

[31] Maxim Pyadushkin and Bradley Perrett, "Russia, China Agree on Long-Range, Wide-body Airliner Partnership," *Aviation Week and Space Technology*, July 12, 2016.

airlines are independent companies, the government can easily create a targeted tariff that can undo any deal. A common historical practice is for the governments to seek an economic trade offset, for example, a supplier role for some component of the aircraft, a regional maintainer role for the aircraft, or a completely different economic endeavor such as investment in resource extraction in the purchasing nation.[32]

### COMAC Shanghai Subsidiaries

Like many top-down Chinese government development initiatives, local governments often play an integral role in policy implementation. Beyond the desire to meet the central authorities' policy goals, local governments seek to generate local economic growth at the provincial and municipal levels. Local economic growth produces local government fees and taxes as well as stimulates real estate prices from which the provincial officials benefit. For example, the municipal government of Shanghai serves a key role in the development of COMAC and of the C919 project. Of COMAC's seven subsidiaries, five are based in Shanghai and are the primary entities behind the development of the C919.

- Shanghai Aircraft Design and Research Institute (SADRI) (上海飞机设计研究院)[33]
  - Based in Zhangjiang Hi-Tech Park, Pudong District, Shanghai, SADRI is primarily responsible for the design, testing, research, and key technologies of the C919 and of most of COMAC's commercial aviation R&D. Before becoming a subsidiary of COMAC, SADRI was the aviation-research arm of AVIC and performed research and design tasks for various civil aircraft programs since its founding in 1970. SADRI completed the design of China's first narrow-body jetliner, the Shanghai Y-10. SADRI also co-designed the Shaanxi Y-8, a medium-sized transport aircraft for military and civilian use.

---

[32] Wesley E. Spreen, *Marketing in the International Aerospace Industry*, Hampshire, UK: Ashgate Publishing, 2007.

[33] Shanghai Aircraft Design and Research Institute, homepage, undated.

- Shanghai Aircraft Manufacturing Company (SAMC) (上海飞机制造有限公司)[34]
  - Established in 1950 and formerly called the Shanghai Aircraft Manufacturing Factory, SAMC is based in Zhangjiang Hi-Tech Park, Pudong District, Shanghai, and is the final assembly and manufacturing center for the C919. SAMC is responsible outsourcing of all assembly parts for the C919 from domestic and international sources and works with COMAC's various R&D centers around China on manufacturing and design of parts. Before becoming a subsidiary of COMAC, SAMC was the manufacturing arm of AVIC and successfully developed China's first narrow-body jetliner, the Shanghai Y-10.[35] SAMC is a joint venture between COMAC and China Reform Holdings Corporation Limited (CRHC). Based in Beijing, CRHC is an asset-management company owned by SASAC that focuses on mergers and acquisitions.[36] CRHC's association with SASAC further substantiates the assessment that the SASAC plays an important role in supporting the development of the C919. CRHC officials have visited COMAC several times to inspect progress of the C919, most recently in May 2016.[37]

- Shanghai Aircraft Customer Service Company (SACS) (上海飞机客户服务有限公司)[38]
  - Based in Shanghai, SACS is COMAC's central service center. It is responsible for training customers and offering aviation materials support and flight training.

---

[34] Shanghai Aircraft Manufacturing Company, homepage, undated.

[35] "Shanghai Aircraft Manufacturing Co.," COMAC Member Organizations, COMAC website, undated.

[36] Bloomberg, "China Reform Holdings Corp Ltd," company profile, undated.

[37] "China Reform Holdings Corporation Liu Dongsheng Visits COMAC [中国国新董事长刘东生到中国商飞访问]," *COMAC News Center*, May 27, 2016.

[38] Shanghai Aircraft Customer Service Co., homepage, undated.

- Shanghai Aviation Industrial (Group) Co. (SAIC) (上海航空工业（集团）有限公司)[39]
  - Based in Shanghai, SAIC is a wholly owned subsidiary of COMAC and is mainly engaged in non-core businesses activities of COMAC, such as asset management, administrative services, and logistic services with civil freight carriers.

- COMAC Shanghai Aviation Flight Test Center (中国商飞民用飞机试飞中心)[40]
  - Based in Pudong District, Shanghai, the COMAC Shanghai Aviation Flight Test Center is responsible for flight tests and safety certification of the C919.

Of these five subsidiaries, SADRI and SAMC represent the most important agencies responsible for the research, development, and testing of the C919. All five benefit from support from the Shanghai municipal government in some form or another, either through direct investments or by receiving preferential tax subsidies by virtue of their location within the Zhangjiang Hi-Tech Park.[41]

**Shanghai Development Banks**

Joint ventures with state development banks in Shanghai have proved an important funding source for COMAC. For example, in 2012, COMAC launched an aviation-finance lease company with Pudong Development Bank (PDB) and Shanghai International Group (SIG) with a reported registered capital of 2.7 billion RMB.[42] PDB is the controlling stakeholder with a 66.67-percent stake, COMAC holds a

---

[39] Shanghai Aviation Industrial (Group) Co., homepage, undated.

[40] COMAC Shanghai Aviation Flight Test Center News Center, homepage, undated.

[41] According to the Zhangjiang Hi-Tech Park website, the park offers a "preferential tax and investment market environment for foreign and domestic businesses." Because of its SEZ status, the park is financially subsidized by the Shanghai municipal government. See Zhangjiang Hi-Tech Park website, undated.

[42] Katie Cantle, "COMAC Establishes Finance Lease Company for C919 Sales," *China Aviation Daily*, May 16, 2012.

22.22-percent stake, and SIG holds an 11.11-percent ownership.[43] The main purpose of the joint venture is to raise capital for COMAC's C919 project, among other COMAC activities. In 2016, PDB and COMAC also signed a "strategic cooperation agreement," in which both sides agreed to cooperate on banking, finance, aircraft leasing, and consulting.[44] In addition to PDB, CDB and its Shanghai affiliates also remain large investors in COMAC.[45]

### Shanghai Municipal Economic and Information Commission

COMAC also leverages its relationship with the Shanghai Municipal Economic and Information Commission (SEIC) to further its development goals. SEIC's role is to implement the laws, rules, and regulations related to the industrial and information needs of the government of Shanghai.[46] SEIC receives guidance from and coordinates policy with the MIIT and the Chinese State Council. SEIC officials frequently visit COMAC's headquarters to inspect progress on the C919 and coordinate efforts with other local government bureaucracies with a stake in the C919 program. For example, in May 2016, a senior SEIC official visited COMAC to tour the C919 final assembly line. SEIC officials were accompanied by officials from the General Office of Inspection and Supervision of the State Council; the State Administration of Science, Technology and Industry for National Defense; and the Shanghai municipal government.[47] During the visit, COMAC officials expressed appreciation for the "long-term support" from the Chinese

---

[43] Cantle, 2012.

[44] "COMAC Signs Strategic Cooperation Agreement with SPD Bank," *COMAC News*, January 28, 2016.

[45] "China Development Bank Financial Leasing Co., Ltd," application proof, *Hong Kong Exchange News*, February 2016.

[46] Shanghai Municipal People's Government, "Shanghai Municipal Commission of Economy and Informatization," website, June 7, 2010.

[47] "General Office of Inspection and Supervision Team Inspects COMAC [国办督查组到中国商飞调研]," *COMAC News Center*, May 30, 2016.

State Council, SEIC, and local government agencies.[48] Officials from SEIC regularly attend events related to the development of the C919, including the founding of SADRI.[49]

### Ties with Other Local Provincial Governments

Finally, COMAC appears to leverage ties with other provincial governments across China. The existence of such ties are in part because of a desire by COMAC officials to broaden the domestic supply chain, manufacturing base, and R&D links with research universities across China. In 2016 alone, provincial and municipal party secretaries from Sichuan,[50] Guangxi,[51] Yunnan,[52] Anhui,[53] Ningxia,[54] and Hong Kong visited COMAC to discuss areas of cooperation.[55]

## Chinese Policies and Laws on Commercial Aviation Joint Ventures

Although no language exists within public Chinese joint-venture law that expressly stipulates preferential treatment or policies for China's

---

[48] "General Office of Inspection and Supervision Team Inspects COMAC [国办督查组到中国商飞调研]," 2016.

[49] "Shanghai Aircraft Design and Research Institute [上海飞机设计研究院成立]," *Sina News*, December 1, 2009.

[50] "He Dongfeng Meets Sichuan Provincial Secretary Yin Li [贺东风拜会四川省省长尹力]," *COMAC News Center*, April 22, 2016.

[51] "Guangxi Party Representative Visits COMAC [广西党政代表团到中国商飞调研]," *COMAC News Center*, May 18, 2016.

[52] "Yunnan Party Representative Visits COMAC [云南省党政代表团走进中国商飞]," *COMAC News Center*, April 20, 2016.

[53] "Wuhu Mayor Pan Chaohui Visits COMAC [芜湖市市长潘朝晖到中国商飞访问]," *COMAC News Center*, May 9, 2016.

[54] "Ningxia Hui Minority Special Administrative Zone Vice Party Secretary Li Rui Visits COMAC [宁夏回族自治区副主席李锐到中国商飞调研]," *COMAC News Center*, May 7, 2016.

[55] "Hong Kong NPC Research Team Visits COMAC [港区全国人大代表专题调研团走进中国商飞]," *COMAC News Center*, May 8, 2016.

commercial aviation industry—such as policies promoting Chinese majority ownership, domestic production agreements, or technology transfer—most joint ventures in practice favor Chinese domestic manufacturers.[56]

For example, there are indications that COMAC's C919 project encouraged, if not required, foreign suppliers of the jetliner components to be manufactured in China or manufactured via joint-venture arrangements. Most of the C919's critical systems, including engines and avionics, are supplied by Western companies or joint ventures between foreign entities and China. A 2013 report from the International Trade Administration found that "most of the major systems procurements" from foreign suppliers were required to be manufactured in mainland China.[57] The one exception, the report found, was the CFM LEAP-X1C engine, developed by CFM International, a joint venture between GE Aviation and French firm Safran, and manufactured in the United States. One Western industry insider remarked that COMAC's insistence that the majority of components of the C919 be manufactured by Chinese suppliers amounted to "giving away [foreign] technology to play on this jet."[58]

Another report from a Chinese-language investigative magazine examining the C919 project found that the degree of "local industrialization" (国产化率) of the jetliner increased from "10 percent in 2008 to 60 percent in 2015," in part because of COMAC requirements that foreign suppliers employ Chinese companies for manufacturing.[59] Tax and import-duty policies on aviation components may have also played

---

[56] Chinese law emphasizes that "joint ventures established within China's territory should be able to promote the development of China's economy and the raising of scientific and technological levels for the benefit of socialist modernization." See "Law of the People's Republic of China on Joint Ventures Using Chinese and Foreign Investment," China.org.cn, March 16, 2007.

[57] Office of Transportation and Machinery, Aerospace Team, Industry Reports, "China (2013)," *International Trade Administration*.

[58] Paul Traynor, "China Just Unveiled Its First Large Passenger Plane" Associated Press, November 2, 2015.

[59] Wang Tao, "We Should Cheer the High Expectations of the C919 [王韬: 应为C919的超预期成果喝彩]," *Observer* [观察者], November 3, 2015.

**Figure 3.3**
**AECC Shareholders and Subsidiaries**

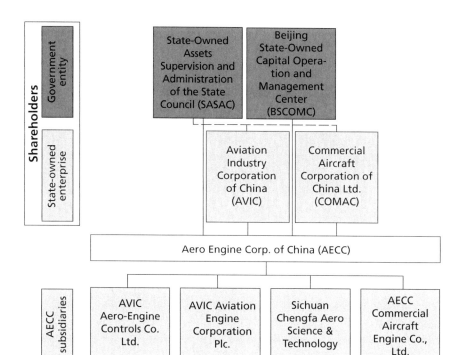

SOURCE: Compiled by authors from multiple sources.
RAND RR1755-3.3

a role, as there are significant import tariffs on aerospace components. For example, in 2014, the Ministry of Finance reduced the tax rate for large aircraft from 17 percent to 5 percent for aircraft imported into China for the purpose of leasing to Chinese airlines.[60] In June 2016, Embraer closed joint venture assembly facilities in Harbin for their regional jets because the government would not extend tariff relief to the new E190 aircraft, presumably to protect ARJ21 domestic sales.[61]

---

[60] U.S.-China Business Council, *USCBC China Economic Reform Scorecard: Steps Forward Undermined by Steps Back*, October 2016, p. 106.

[61] Mavis Toh, "Embraer Could Cease Production at Harbin," *FlightGlobal*, September 7, 2015; Bradley Perrett, "Embraer, Avic Will Close Joint Harbin Company," *Aviation Week*

COMAC, for its part, does not publish public-tender announcements or policies that explicitly mandate that all foreign suppliers must use Chinese manufacturers. COMAC executives do claim, however, that "all the major parts of the (C919) plane were designed, tested, and manufactured in China," an inaccurate but nonetheless powerful expression of Chinese prowess and pride.[62]

### Aero Engine Corporation of China

In August 2016, the Aero Engine Corporation of China (AECC) was inaugurated. While AECC is primarily a spin off of AVIC's engine divisions, the goal appears to focus resources and attention on producing more advanced aircraft engines for both commercial and military platforms.[63] This reorganization has been in the works for a number of years. All the AECC subsidiaries previously existed; the only real change is to the ownership structure pictured in Figure 3.3 from primary AVIC control. The new AECC structure has aspects of a COMAC subsidiary, a JV with AVIC, and a tie with the Beijing provincial government, making it more than a simple domestic engine supplier.

The impact of the management change will take time to be evident as engine product cycles take many years. However, what is evident today is the similarity to the COMAC structure previously discussed in detail. China's state shareholders' attention, commercial aviation industrial policies, and associated resources can be focused more directly on AECC. Although all were applicable to AVIC, AVIC has a much broader range of business interests including military aviation, automotive industry, electronics, and others.

## Findings on China's Government Policy

- China adopts a systematic, whole-of-government approach to developing its domestic commercial aviation industry. The pro-

---

*and Space Technology,* June 4, 2016a.

[62] Yin Pumin, "C919, Made in China," *Beijing Review,* November 19, 2015.

[63] Perrett, 2016b.

cess starts with broad executive-level policy guidelines issued by the PRC State Council, followed by more detailed ministerial-level policy planning documents. These Chinese ministries then coordinate with local government and SOE networks to build commercial aviation conglomerates such as COMAC. COMAC has developed deep business and financial networks in Shanghai and within China's regional and national science and technology bureaucracy. Such ties are manifested in financial investments in COMAC as well as in meetings between senior executives of COMAC and leaders of industry and government in Shanghai and beyond. The C919 jetliner project has been funded in large part because of assistance from the Central and Shanghai governments, their development banks, SOEs, and commissions.

- **China's commercial aviation industry is unlikely to reach its goals.** The ARJ 21 is being delivered only to COMAC's Chengdu Airlines and has not yet received FAA certification and the first flight of the C919 has been delayed until 2017. Even when the C919 completes testing and if it gets FAA certification, it is not clear whether the aircraft has the efficiency or global supply chains to compete with Boeing and Airbus even within China. While China's policies and tariffs on aircraft may compel their domestic purchase and operation, operating these aircraft will likely cost more per seat, making the airline less competitive with those operating more-efficient aircraft or high-speed trains. The ultimate metric of success would be the export of Chinese aircraft, their operation by foreign airlines, and a growing backlog of aircraft orders from foreign airlines without subsidies from the Chinese government.

Chapter Four looks at Chinese investment in U.S. aviation and aviation-based relationships between Chinese institutions and U.S. universities. While Chinese investors continue to make more significant investments in U.S. aviation than in the past, it does not appear to be related to China's push to be competitive in the global LCA market.

# Chinese Investments in U.S. Aviation

This chapter documents Chinese investments in U.S. aviation since 2006. Those investments have taken a variety of forms. They range from the obvious Chinese acquisition of or merger with a U.S. aerospace company that manufactures aircraft to many less-obvious acquisitions, such as U.S. manufacturing companies (part of aviation or other industry supply chains), to joint ventures between Chinese and U.S. aviation companies to partnering agreements between Chinese and U.S. aviation companies. There are also a few publicly known failed aviation-investment deals among parties in the two countries.

All the investments we identified are listed in Figures 4.1 and 4.2 and briefly described. The investments are relatively small and mostly in GA companies. In Chapter Five, we discuss the strategy behind these investments. Finally, we also assessed relationships between Chinese entities and U.S. university programs relevant to aviation. In general, we found very few aviation-based relationships between U.S. educational institutions and China.

## Descriptions of Chinese Investments in U.S. Aviation

### Mergers and Acquisitions

We identified 12 mergers or acquisitions of either U.S. aviation companies by Chinese entities or U.S. industrial companies by Chinese aviation companies. Most are producers of light aircraft, such as helicopters and GA airplanes or aviation parts. To the best of our knowledge, all of these mergers and acquisitions are less than $1 billion. If Henniges

**Figure 4.1**
**Timeline of Chinese Investments in U.S. Aviation**

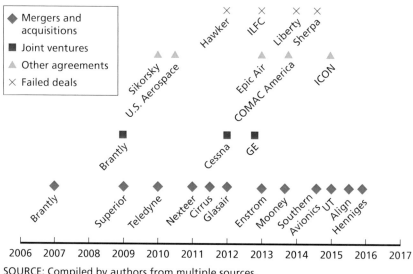

SOURCE: Compiled by authors from multiple sources.
RAND RR1755-4.1

and Nexteer were eliminated as pure automotive companies, then the largest deal we can document is the Cirrus Aircraft acquisition at just more than $200 million. These are all small GA companies with little effect on U.S. competitiveness or national security.

**Brantly International Limited** is a light helicopter manufacturer that Qingdao Haili Helicopters acquired in 2007. The latter is a helicopter manufacturer based in Shandong and chaired by Cheng Shenzong.[1] Referred to in English-language media as the "helicopter king," Cheng is a Chinese businessman with a reputation for being a complete neophyte in the aviation industry.[2] Although Cheng's investment partners include some Beijing provincial financing entities, it

---

[1]   Bradley Perrett, "Chinese Bizjet Mismatch: Demand vs. Assembly Plans," *Aviation Week and Space Technology*, October 14, 2013.

[2]   "The 'Helicopter King of China' Is Quietly Building an Empire," *Business Insider*, July 13, 2012; Zhu Qionghua [朱琼华], "'Acquisition Maniac' Cheng Shenzong: 'Neophyte' and 'Hyper' [并购狂人成身棕: '外行'与'炒炸者']," *Phoenix Finance* [凤凰财经], July 19, 2012.

**Figure 4.2**
**Value of Chinese Investments in U.S. Aviation**

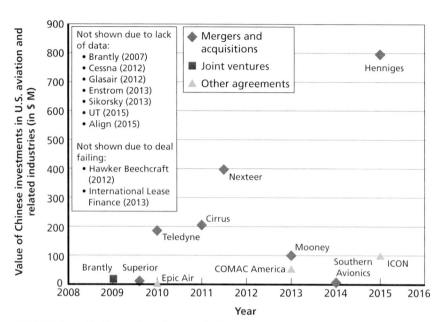

SOURCE: Compiled by authors from multiple sources.
RAND RR1755-4.2

is unclear how he finances these deals. One unnamed Chinese aviation industry executive stated of Cheng, "[He] is very enthusiastic but
does not understand aviation. He has absolutely no direction when
doing things and has no success stories. . . . The companies that he
has acquired overseas are nearly all useless."[3] Under Qingdao's ownership, Brantly then went on to acquire Superior Air Parts in 2009.[4] In
2011, Brantly relocated its engineering and administrative offices from
Vernon, Texas, to Superior Air Parts' facilities in Coppell, Texas.[5]

---

[3]  Zhu Qionghua, 2012.

[4]  United States Bankruptcy Court for the Northern District of Texas, Dallas Division,
"Memorandum Opinion and Order Denying Motion to Enforce for Lack of Subject Matter
Jurisdiction," August 20, 2014.

[5]  Brantly B-2B Helicopter, "Service Bulletin 111," website, February 28, 2011.

**Superior Air Parts** was a "U.S. manufacturer of aftermarket parts for Continental and Lycoming piston engines."[6] They provided replacement parts and accessories for aircraft piston engines from other manufacturers. Following the 2009 bankruptcy of the company's parent, Thielert, a German holding company, Brantly/Weifang Tianxiang Technology Group acquired it and named the new entity Superior Aviation.[7] In 2012, Superior Aviation attempted to acquire Hawker Beechcraft, but the deal ultimately failed because of concern that the Committee on Foreign Investment in the United States (CFIUS) would not approve it.[8]

**Teledyne Technologies Continental Motors and Teledyne Mattituck Services** constituted the GA piston engine business of Teledyne Technologies Incorporated. On December 24, 2010, Teledyne announced an agreement with AVIC International Holding Corporation to sell them both to Technify Motor, a subsidiary of AVIC International, for $186 million in cash.[9] The Technify Motor/AVIC subsidiary, Continental Motors, later acquired United Turbine and UT Aeroparts Corporations in January 2015.[10]

**Nexteer Automotive** is a "Michigan-based maker of steering and driveline systems."[11] According to Nexteer's website, "PCM China, a company controlled by Beijing E-Town, the financing and investment arm of the Beijing Municipal Government," acquired it in December 2010 from General Motors. AVIC Auto, a subsidiary of AVIC, subse-

---

[6]  Perrett, 2013.

[7]  United States Bankruptcy Court for the Northern District of Texas, Dallas Division, 2014; and Mark Phelps, "Superior Air Parts Is Back from Bankruptcy," *Flying*, August 5, 2010.

[8]  Mike Spector, "Hawker Sales Talks Collapse over Review Worries," *Wall Street Journal*, October 18, 2012.

[9]  "Teledyne Technologies Agrees to Sell Teledyne Continental Motors to AVIC International," Teledyne Technologies, December 14, 2010.

[10] "Continental Motors Services Acquires United Turbine and UT Aeroparts Corporations," Continental Motors, February 2, 2015.

[11] Han Tianyang, "State-Owned AVIC Buys US-Based Nexteer," *China Daily*, April 11, 2011.

quently acquired a 51-percent equity in PCM China, thereby becoming the controlling shareholder of Nexteer in March 2011.[12] At the time it was acquired, "Nexteer [was] the world's third-largest company in sales of driveshaft components and the fourth-biggest for steering systems."[13] The transaction is included due to its connection with AVIC, but it remains an automotive component supply company.

**Cirrus Aircraft** is one of the world's largest makers of GA aircraft, second only to Cessna, and is also the world's largest maker of piston-powered GA aircraft.[14] Cirrus announced a deal in February 2011 that would give the Chinese Aviation Industry General Aircraft (CAIGA, a subsidiary of AVIC) 100-percent ownership of Cirrus, and the merger was completed four months later in June 2011.[15] According to *Flying Magazine*, "CAIGA [acquired] Cirrus Aircraft from a Bahraini investment group, which [had] owned the company since 2001."[16] The U.S. Chamber of Commerce also noted that "[t]he purchase was the first acquisition of an aircraft maker by a Chinese company in the United States and the third acquisition by AVIC in the United States."[17]

**Glasair** is a kitplane maker based in Arlington in Washington State. On July 20, 2012, it reached an agreement with Jilin Hanxing Group, whereby Jilin Hanxing acquired Glasair's manufacturing assets. Jilin Hanxing itself appears to be a private corporate group primarily involved in "travel, culture, energy, logistics, chemical, real estate development, automobile service, [and] general aviation." The group's chair, Fang Tieji, is a member of the National People's Congress, representing Jilin Province, but the company's website does not

---

[12] Nexteer Automotive, "History," website, undated; and Han Tianyang, 2011.

[13] Han Tianyang, 2011.

[14] Yu Dawei [于达维], "CAIGA Acquires U.S. Plane Manufacturer [中航工业通飞并购美国飞机制造商]," *Caixin Online* [财新网], March 1, 2011.

[15] Robert Goyer, "New Owners for Cirrus," *Flying*, February 28, 2011; and Bethany Whitfield, "Cirrus Completes Merger with Chinese Firm CAIGA," *Flying*, June 29, 2011.

[16] Whitfield, 2011.

[17] U.S. Chamber of Commerce, *Faces of Chinese Investment in the United States*, 2012, p. 10.

indicate any connection with AVIC or any other government entity.[18] On November 14, 2012, Jilin Hanxing acquired Glasair and renamed it the Zhuhai Hanxing General Aviation Company.[19] Jilin Hanxing appears also to be investing in GA infrastructure along the east coast of China, perhaps in synergy with its other lines of business in travel, logistics, and energy.[20]

**Enstrom Helicopter Corporation** is a maker of light helicopters. On January 4, 2013, Chongqing Helicopter Investment Company acquired 100 percent of the stock rights of Enstrom. According to a press release on its website, this was the first time that a Chinese aviation company successfully acquired a well-known helicopter manufacturing company in either the United States or Europe.[21] It should be noted that Brantly International, another light helicopter manufacturer, was acquired in 2007, although Enstrom has more production lines and a larger global footprint.

**Mooney Aviation Company** is a manufacturer of high-performance piston-engine aircraft based in the United States. Meijing Group, a Chinese real estate developer, acquired it in October 2013 for about $100 million with a promise to invest another $1 billion at a later stage. CFIUS approved the deal on October 2, 2013.[22] The company resumed production in 2014 after entering a self-imposed hibernation in 2010.[23] According to media reports, the investment by Meijing was "apparently sufficient" to allow Mooney to restart productions.[24] Mooney remains in operation today, and on August 16, 2016, the company appointed a new president and chief executive officer.[25]

---

[18] Jilin Hanxing Group, "Group Profile," website, undated.

[19] Jilin Hanxing Group, "Events [大事记]," website, undated.

[20] "Company Profile: Hanxing Group," undated.

[21] Chongqing Helicopter Manufacturing Investment Company, "CQHIC Acquires U.S. Enstrom [重庆直投收购美国恩斯特龙]," website, January 5, 2013.

[22] "Chinese Firm Completes U.S. Aircraft Maker Merger," *China Daily*, October 17, 2013.

[23] "Mooney 'Hibernation' Ends, Texas Factory Is Humming," AOPA, May 5, 2015.

[24] Stephen Pope, "Mooney Aviation Back in Business," *Flying*, October 15, 2013.

[25] "Mooney International Appoints New President and CEO," *Mooney*, August 16, 2016.

**Southern Avionics** provides avionics services and interior upgrades for small aircraft. In late 2014, it was acquired by aforementioned Continental Motors, an AVIC subsidiary. It was a small company of 14 employees.[26]

**United Turbine and UT Aeroparts Corporations** specialize in Pratt and Whitney PT6 overhaul, repair, and parts sales. Continental Motors Group, a subsidiary of AVIC, acquired them on January 31, 2015.[27]

**Align Aerospace** "provides supply chain services for the aerospace industry and is a leading global distributor of fasteners and other hardware, primarily to aerospace original equipment manufacturers and suppliers."[28] AVIC International Holding (Zhuhai), a subsidiary of AVIC, acquired it on March 31, 2015.

**Henniges Automotive Holdings** is "a leading supplier for the global automotive market of so-called dynamic sealing and anti-vibration solutions that keep vehicles quiet and dry." AVIC Auto and U.S.-China investment firm BHR acquired it for $600 million in 2015, "the biggest Chinese investment in U.S. automotive manufacturing assets to date."[29] BHR is the "cross-border investment platform of Bohai Industrial Investment Fund (the first RMB private equity fund approved by [the PRC] State Council),"[30] which, in turn, is controlled by the Bank of China, one of the largest state-owned banks in China.

---

[26] Kerry Lynch, "Continental Motors Adding Avionics to Expanding Capabilities," *Aviation Week Intelligence Network*, June 24, 2014.

[27] Continental Motors, "Continental Motors Services Acquires United Turbine and UT Aeroparts Corporations," February 2, 2015.

[28] "AVIC International Expands Commercial Aerospace Services Portfolio with the Acquisition of Align Aerospace," AVIC International website, March 31, 2015.

[29] Bien Perez, "Chinese Direct Investment in U.S. to Top US$10 Billion for Third Year in a Row," *South China Morning Post*, November 13, 2015.

[30] "BHR and AVIC Auto Acquire Henniges Automotive," PR Newswire, September 15, 2015.

## Joint Ventures

The four joint ventures appear to facilitate doing business between the U.S. and Chinese companies. As of 2016, none of them appears to have been a significant commercial success.

In 2009, **Brantly**, which was owned by Qingdao Hail Helicopters, established Weifang Tianxiang Aerospace Industry Company Limited as a joint venture with Weifang Tianxiang Aerospace Technology Company. Cheng Shenzong also chairs the latter company. The joint venture had an investment of $20 million and registered capital of $80 million. The joint venture's primary product would be Brantly's light helicopter.[31]

While Brantly's light helicopter does not appear to be in production, by 2011, Weifang Tianxiang had developed a remote-controlled helicopter supposedly based on the Brantly design.[32] Reports in 2016 suggest that the resulting unmanned autonomous vehicle is being militarized.[33] The Brantly intellectual property was not advanced technology, but an inexpensive platform that first flew in the 1950s to which Weifang Tianxiang added automation and potentially weapons.

On November 16, 2009, **GE Aviation** of the United States and AVIC "announced [an] agreement on forming a new joint venture company to develop and market integrated avionics systems for commercial aircraft customers." The new company, headquartered in China, intends to "develop and market avionics systems for commercial aircraft customers."[34] According to the press release announcing the agreement, it would supply to airframers worldwide and act as a medium through which GE and AVIC could jointly bid solutions to

---

[31] "The 'Helicopter King of China' Is Quietly Building an Empire," *Business Insider*, July 13, 2012; and "Shandong and U.S. Cooperate to Produce First Light Helicopter Next Year [山东与美国合作明年将生产出首架轻型直升机]," *Sina*, September 3, 2009.

[32] "China Succeeds in Its Largest Unmanned Helicopter's First Flight," *Global Times*, May 8, 2011.

[33] Jeffrey Lin and P. W. Singer, "China's Armed Robot Helicopter Takes Flight," *Popular Science*, July, 11, 2016.

[34] "GE and AVIC Joint Venture Creates New Global Business Opportunities," GE Aviation website, November 16, 2009.

compete for the C919 program, which would include an "open architecture Integrated Modular Avionics [IMA] platform."[35]

On January 21, 2011, GE and AVIC finalized the formation of the joint venture company. The new company works to "develop and market integrated, open-architecture avionics systems to the global commercial aerospace industry for new aircraft platforms . . . [that] will host the airplane's avionics, maintenance and utility functions."[36] According to the press release announcing this joint venture, it would be known as GE-AVIC Civil Avionics Systems Company Limited, and its initial focus would be "integrated avionics systems for the C919 aircraft."[37] After nearly two more years of additional preparatory work, the joint venture was finally launched on October 22, 2012, out of Shanghai.[38] It appears that, through 2016, GE-AVIC joint ventures have focused on the C919 exclusively because of the lack of other reported business deals.

**Cessna**, the world's largest maker of GA aircraft, announced in March 2012 that it had signed two deals—one with AVIC, the other with AVIC and the Chengdu government—to produce midsize business jets and other aircraft in China.[39] That same year, Cessna also signed two agreements with CAIGA to establish two joint venture companies: one based in Zhuhai and another based in Shijiazhuang.[40] However, in the case of Shijiazhuang, Cessna clarified that its Caravan models would continue to be manufactured in Kansas before being shipped to China for final assembly.[41] From 2009 to 2012, Cessna

---

[35] "GE and AVIC Joint Venture Creates New Global Business Opportunities," 2009.

[36] "GE and AVIC Sign Agreement for Integrated Avionics Joint Venture," GE Aviation website, January 21, 2011.

[37] "GE and AVIC Sign Agreement for Integrated Avionics Joint Venture," 2011.

[38] Katie Cantle, "AVIC, GE Aviation Formally Launch Integrated Avionics JV," *Air Transport World*, October 22, 2012.

[39] Bethany Whitfield, "Cessna Signs Deal to Build Jets, Other Aircraft in China," *Flying*, March 27, 2012.

[40] "Cessna and CAIGA Joint Venture to Start Operations [赛斯纳和中航通用飞机的合资企业即将运营]," *China Daily* [中国日报], April 16, 2013.

[41] Stephen Pope, "Cessna to Assemble Caravans in China," *Flying*, May 3, 2012.

manufactured the Cessna 162 Skycatcher light sport aircraft in China by Shenyang Aircraft Company, a subsidiary of AVIC, presumably to reduce cost, but that plan was canceled because of failure to meet cost goals and lack of orders.[42] By 2014, Cessna had overall "scaled backed its Chinese program to only the Citation XLS+, which was the smallest of the two business jets, and the Caravan."[43] It had further dropped AVIC Aviation Techniques as one of its partners and concentrated production to just one site because of production costs. The customers for these products would be Chinese, and they do not appear to be exporting them from China.[44]

### Other Types of Deals

Finally, there are five additional business deals that do not fit neatly in the other categories.

**Epic Air** was a kit airplane manufacturer that filed for bankruptcy on September 10, 2009. Following approval by a bankruptcy judge on April 12, 2010, CAIGA purchased Epic Air's assets for $4.3 million. The asset purchase agreement did not include "any defense-related material that might be subject to International Traffic in Arms regulations."[45] At the time of the agreement, CAIGA had signed a memorandum of understanding with LT Builders Group, another bidder, whereby the former would license to the latter "the intellectual property and technology for the aircraft known as the 'EPIC LT.'"[46] LT Builders owned all of Epic's design rights. It subsequently licensed these design rights to CAIGA, which also "obtained the rights for sales of experimental aircraft in other markets and obtained some aircraft tooling excluding that for the Epic LT." On March 6, 2012, Epic Air-

---

[42] "Cessna and CAIGA Joint Venture to Start Operations," 2013; John Zimmerman, "The Skycatcher's Death Proves the LSA Rule Is a Failure," *Air Facts*, April 21, 2014.

[43] Bradley Perrett, "Cessna Downsizes Its Chinese Assembly Plans," *Aviation Week*, April 7, 2014.

[44] Perrett, 2014.

[45] Matt Thurber, "Chinese Firm to Buy Epic Assets," AIN online, April 30, 2010.

[46] Thurber, 2010.

craft was sold to a Russian maintenance, repair, and overhaul company called Engineering LLC in a deal whose terms were not disclosed.[47]

**Sikorsky Aircraft** signed a deal with Changhe Aircraft Industries Corporation on September 5, 2013, for Changhe to produce S-76D commercial helicopter cabins for Sikorsky. Changhe previously provided cabins for the S-76C++ helicopter under an agreement signed in 2007.[48]

**ICON Aircraft** is a Los Angeles–based company specializing in designing and manufacturing light sport aircraft, including its A5 twin-seat amphibious plane. In July 2015, it signed a deal with Shanghai Harbor City Development Company Limited and Shanghai Pudong Science and Technology Investment Company Limited to create a China headquarters in Pudong. Shanghai Pudong Science and Technology Investment Company also agreed to invest $100 million in ICON "to become its largest shareholder and help ICON find production partners in China."[49] As of late 2016, ICON Aircraft appears to still be headquartered in California.

**U.S. Aerospace Inc.** was a small California-based company that teamed up with AVIC in February 2011 to "try to launch bids for U.S. defense contracts, possibly including one to supply Chinese helicopters to replace the aging Marine One fleet used by the president."[50] U.S. Aerospace Inc. was a defense subcontractor that supplied "aircraft assemblies, structural components and highly engineered, precision-machined details for commercial and military aircraft."[51] U.S. Aerospace Inc. is no longer an ongoing business with no listed stock price or web presence. None of these ventures with AVIC appears to have come to fruition.

---

[47] Mark Huber, "Epic Sold to Russian MRO," AIN online, April 2, 2012.

[48] "Sikorsky and Changhe Sign Agreement for S-76D Cabin Production in China," Sikorsky, September 5, 2013.

[49] Pudong New Area Government, "Light Sport Aircraft to Be Produced in Pudong," e-Pudong, July 13, 2015.

[50] Jeremy Page, "China Eyes U.S. Defense Contracts," *Wall Street Journal*, February 4, 2011.

[51] "U.S. Aerospace, Inc. and AVIC International Holding Corporation Enter into Strategic Cooperation Agreement," *Business Wire*, September 20, 2010.

**COMAC America** was established in California as a wholly owned subsidiary of COMAC in 2013 with a commitment to invest $50 million. It does not appear to have operations independent of COMAC.[52]

### Failed Deals

Investors and the press publicly discussed the four failed deals. Confidential discussions may have been held or even pursued for other investments that ultimately failed, possibly because of concerns of CFIUS denials or withholding of export approvals.

CFIUS is a U.S. government interagency committee that reviews corporate transactions involving foreigners for national security concerns. However, because of confidentiality of CFIUS proceedings, it is unclear how these potential deals may have been reviewed by the U.S. government. U.S. presidents have required divestitures of two transactions involving Chinese firms: MAMCO Manufacturing as a whole in 1990 and Ralls Corporation wind farms in 2012. Export approvals are conducted by the U.S. Department of State or the U.S. Department of Commerce with input from other government departments for national security concerns to the movement of goods or arms. The process is similarly confidential and may have altered Chinese investment plans.

**Hawker Beechcraft** was a corporate jet and piston-engined plane maker that filed bankruptcy in 2012. That same summer, Superior Aviation Beijing Company attempted to buy these operations out of bankruptcy for $1.79 billion. According to the *Wall Street Journal*, Superior "encountered difficulties separating the Wichita, Kansas–based company's defense business from those units in a way that would make both sides comfortable the deal would get U.S. government clearance."[53] The deal ultimately collapsed in October 2012.[54]

---

[52] Wang Jun, "COMAC's First U.S. Subsidiary Set to Take Off," *China Daily*, November 26, 2013.

[53] Spector, 2012.

[54] Spector, 2012.

**International Lease Finance Corporation (ILFC)** was the wholly owned aircraft-leasing arm of American International Group (AIG). A consortium of Chinese firms—comprising New China Trust, P3 Investment, and China Aviation Industrial Fund—attempted to acquire it for $4.3 billion.[55] CFIUS granted initial approval on June 1, 2013. However, the consortium ultimately failed to put together the necessary financing, and AIG agreed to sell ILFC to AerCap Holdings for $5.4 billion in stock and cash in December 2013.[56]

Based in Scappoose, Oregon, **Sherpa Aircraft** makes the Sherpa 650T bush plane. Bush planes are used to fly to remote areas that often do not have good infrastructure.[57] In January 2013, it had secured funding from a Chinese investor to support 24 months of certification. It also formed a joint venture called Ying-Kou Sherpa on January 4, 2013. However, the funding deal collapsed by April 2014.[58] Sherpa Aircraft remains in operation and, on July 21, 2016, it released a new Model 600 plane.[59]

**Liberty Aircraft** is a company that had signed a deal with an unnamed Chinese investor to "build the two-seat trainer, designed by the same engineer who built the British European kitplane, in China." However, the deal collapsed before or around April 2014 for undisclosed reasons.[60]

---

[55] Charlotte So, "Chinese Consortium Cleared for ILFC Deal," *South China Morning Post*, June 1, 2013.

[56] Michael J. de la Merced, "A.I.G. Sells Aircraft Leasing Unit for $5.4 Billion," *New York Times*, December 15, 2013.

[57] "Best Bush Planes: Flying Cessna, Piper, Beech, DeHavilland, Airplanes and Aircraft," Bush-planes.com, undated.

[58] "Sherpa Strikes China Deal for Certification," AOPA, February 7, 2013; and "Sherpa Aircraft Deal with Chinese Investor Fails," AOPA, April 3, 2014.

[59] "News," Sherpa Aircraft, July 21, 2016.

[60] "Sherpa Aircraft Deal with Chinese Investor Fails," 2014.

## Investment Patterns

Despite having essentially no previous history of investing in U.S. aviation, China has acquired or invested in more than a dozen U.S. aviation companies in the past decade. Since 2005, Chinese companies have steadily increased investment in U.S. aviation by acquiring, merging, or establishing joint ventures with U.S. aviation companies without directly running afoul of U.S. regulation, especially when taking into account the size of the known failed deals. However, the deals that have passed muster remain relatively small GA companies with little effect on U.S. competitiveness or national security. As the U.S. Chamber of Commerce notes, CAIGA's purchase of Cirrus in 2011 was the first acquisition of a U.S. aircraft maker by a Chinese company.

Chinese government entities were also responsible for roughly half of the successful deals listed in this chapter. AVIC, responsible for most of those, has undoubtedly been the most prolific investor in U.S. aviation. It often acted through subsidiaries, such as CAIGA, or has acquired a U.S. company and then used that company to acquire others. AVIC's investments also extended outside of the aviation industry to include the automotive industry, as AVIC also produces buses, specialty vehicles, and automotive parts. AVIC has also invested in such U.S. nonaviation/nonindustrial businesses as hotels as well as foreign companies with U.S. operations. These include AVIC International Shenzhen's March 2013 acquisition of DoubleTree by Hilton San Pedro for $12 million; AVIC USA's September 2013 acquisition of three hotels in Fullerton, California; Atlanta, Georgia; Southfield, Michigan; and AVIC Electromechanical and other Chinese investors' May 2014 acquisition of Hilite International, a German company with U.S. operations, for about $250 million.[61]

The private businessman Cheng Shenzong was the second most prolific investor, acquiring several U.S. companies through various Chinese companies that he chairs. He was responsible for three successful deals and one failed deal. He also used the same techniques as

---

[61] Thilo Hanemann and Daniel H. Rosen, *Chinese Investment in the United States: Recent Trends and the Policy Agenda*, New York: Rhodium Group, December 2016.

AVIC did, first acquiring an American company, Brantly, and then using that company to acquire another, Superior Air Parts. While not a large SOE, he still depends on the Chinese banking system which is dominated by state-owned enterprises and government investment funds and associated government policy.

China's accelerating investments in the U.S. aviation industry notwithstanding, the significance of these activities is questionable. These deals were almost all small scale, each worth less than $500 million, with only three exceptions. Of these three exceptions, only one was successful: AVIC Auto's acquisition of Henniges for $800 million in 2015, which, despite its association with AVIC, is really automotive in nature. The two overall biggest deals failed, however. Superior Aviation, chaired by Cheng Shenzong, attempted to acquire Hawker Beechcraft for $1.79 billion, but the deal ultimately fell through in 2012. A consortium of Chinese companies, meanwhile, attempted to acquire ILFC from AIG for $4.3 billion, but the deal also fell through in 2013 when the consortium failed to put together enough money.

The nature of these deals further puts their significance into question. Deals ranged from acquisitions of auto-part makers (Nexteer, Henniges) to acquisitions of GA companies (Cirrus) to production agreements (Sikorsky and Changhe Aircraft). None of these deals involves advanced aerospace technologies, let alone military-critical technologies. The business niche for personal aircraft from a Cirrus or Cessna is about price point; their focus is on integrating affordable fiberglass structures and piston engines or cheaper small jet engines into a safe aircraft, and such technologies are as broadly available as automotive technology. The most advanced technology involved in these deals is the GE-AVIC joint venture on avionics, which is tailored to avoid inadvertent technology transfer and based on avionics technology originating in the United Kingdom. There did not appear to be any particular pattern to these deals nor any systematic targeting of strategic companies and sectors. Some of the acquisitions and attempted acquisitions were businesses in distress or bankruptcy. It is not clear that suggests anything other than a potential turnaround situation or costly acquisition for the investor. Given the small sample size of less than two dozen transactions over a decade, it is hard to extrapolate further on the goals of Chinese investment in U.S. aviation. In Chapter Five, through the broader lens of subject-matter experts and the global aviation industry, we further discuss possible Chinese strategies and the implications to the U.S. aviation industry, especially with respect to avionics and engines and associated joint ventures.

## Relationships with U.S. Aerospace Engineering Universities

We assessed Chinese investment in educational relationships with U.S. universities related to aviation by looking for significant public connections between U.S. universities with highly regarded aerospace programs and Chinese entities—government, university, or corporate. Common across many U.S. universities, including those with aerospace programs, is the presence of Chinese nationals as students via study-abroad and Mandarin-language programs. Instead, we looked for Chinese investments in joint efforts (e.g., research and development, laboratories, or degree programs) related to aviation or aerospace where the Chinese aviation industry might benefit through either leveraging U.S. academic talent or facilitating technology transfer. We only found a handful of such efforts related to aerospace. For completeness, we also describe other non-aerospace connections or the lack of connections found at universities with top-ranked aerospace programs. As a comparative point, we looked to see if there have been any relationships similar to the £3 million (roughly $4.5 million in 2012) research deal signed between the University of Nottingham in the United Kingdom and AVIC or like the relationships between other British universities and AVIC, which are now under investigation by U.S. authorities.[62] We did not find any evidence of such relationships.

---

[62] "The University of Nottingham Signs £3 Million Deal with AVIC, One of China's Biggest Aerospace Businesses," University of Nottingham, undated; and Charles Clover, "UK Universities Under Scrutiny over China Ties," *Financial Times*, June 23, 2015.

We looked at the top 35 U.S. aerospace engineering schools, based on a number of online rankings,[63] for any relationships with Chinese entities. Among the 35 schools, only four had significant connections with Chinese entities. Another 18 schools had some connections, which are documented in this section, but these were not related to aerospace engineering or aviation. Another 13 universities had very

---

[63] We started with *U.S. News and World Report*'s best undergraduate and graduate aerospace programs from the past couple of years, which provides a little more than a dozen programs. Then, we cross-referenced a larger list of U.S. aerospace programs with *U.S News and World Report*'s top 50 engineering schools. All but one of the 35 top aerospace engineering schools rank among the top 50 best engineering schools in the United States. We assess that any effect on U.S. aerospace engineering and the U.S. aviation industry from relationships with Chinese entities would be evident from a review of this sample.

**Figure 4.3**
**Timeline of U.S. Universities' Relationship with Chinese Entities**

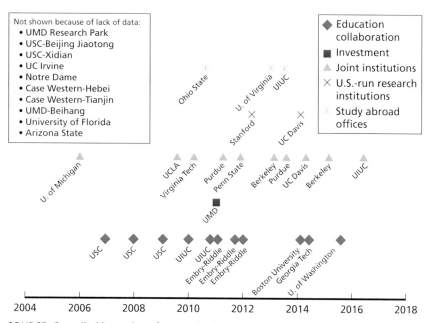

SOURCE: Compiled by authors from multiple sources.
NOTE: UC Davis = University of California, Davis; UCLA = University of California, Los Angeles; UIUC = University of Illinois Urbana-Champaign;  UMD = University of Maryland; USC = University of Southern California.
RAND RR1755-4.3

Table 4.1
Top U.S. Aerospace Engineering Schools with Limited Relationships with Chinese Entities

| School | Rank Among Top Schools | | |
|---|---|---|---|
| | Engineering | Aerospace engineering undergrad) | Aerospace engineering (graduate) |
| Massachusetts Institute of Technology | 1 | 1 | 1 |
| California Institute of Technology (Caltech) | 4 | 5 | 4 |
| University of Texas at Austin | 10 | 10 | 10 |
| Texas A&M University | 11 | 12 | 10 |
| Cornell University | 12 | N/A | N/A |
| University of California, San Diego | 17 | N/A | N/A |
| Princeton University | 18 | 7 | N/A |
| North Carolina State University | 27 | N/A | N/A |
| University of Minnesota | 27 | N/A | N/A |
| University of Colorado at Boulder | 33 | 10 | 8 |
| Rensselaer Polytechnic Institute | 39 | N/A | N/A |
| New York University Tandon School of Engineering | 46 | N/A | N/A |
| Iowa State University | 48 | N/A | N/A |

SOURCE: "Education and Advice," *U.S. News and World Report*, undated.
NOTE: *U.S. News and World Report* for rankings and authors' assessment of limited relationships based on lack of publicly acknowledged relationship.

limited relationships with Chinese entities. Figure 4.3 provides a timeline for the initiation of these relationships. There does not appear to be a systematic attempt by the Chinese government to develop strategically useful relationships with aerospace schools in the United States. Aviation-technology transfer through educational partnerships between U.S. universities and Chinese entities is likely limited to Chinese students enrolled in any of the programs.

The only four U.S. universities that had aerospace-related connections with Chinese entities are the University of Maryland, the University of Michigan, Purdue University, and Embry-Riddle Aeronautical University. Most interesting is the University of Maryland, as AVIC was at one point a member of the consortium supporting the university's Center for Advanced Life Cycle Engineering (CALCE) and continues to maintain informal connections, as detailed below.

Almost half the universities had some broader connection with Chinese entities, but none appeared consequential. Chinese-government entities were parties to few of the relationships. The top partners for U.S. universities were Peking University, Tsinghua University, and Zhejiang University; each was involved in three partnerships. Most of the relationships, however, have nothing to do with aerospace. The relationships included joint institutes focused on non-aerospace engineering, research collaboration on non-aerospace–related topics, and universities establishing satellite offices in China to manage their study-abroad programs.

The remaining 13 schools listed in Table 4.1 essentially have no relationships with any Chinese universities and entities. For Texas A&M University, North Carolina State University, and the University of Minnesota, the limit of their involvement with Chinese entities is hosting Confucius Institutes on their campuses. Specifics on the various connections are provided in the next section.

## U.S. Universities' Relationship with Chinese Universities and Entities

As mentioned, we started with the top 35 aerospace engineering programs in the United States and searched open sources for connections between those universities and Chinese institutions. We list our findings from most significant to least significant with respect to aviation. We did not contact most of the schools nor did we try to ascertain the number of Chinese nationals studying at those universities or in their aerospace programs.

### Aerospace-Related Connections

Embry-Riddle Aeronautical University, ranked 122nd among U.S. engineering schools, established a joint educational program with the

Civil Aviation University of China in October 2011.[64] In November of that same year, it also "entered into agreements to create collaborative degree programs with the University of Shanghai for Science and Technology (USST) and Nanjing University of Aeronautics and Astronautics-Jincheng College (NUAA-JC)." The agreement with USST covers (1) computer, software, and electrical engineering, (2) mechanical engineering; and (3) business. Students in these programs will receive a bachelor of science degree from their home university after three years and a master's degree in science from Embry-Riddle after their fifth year. The agreement with NUAA-JC, meanwhile, creates a collaborative bachelor of science/master of science degree program.[65]

The University of Maryland, College Park, ranked 24th among U.S. engineering schools, has relationships with a number of Chinese entities. China Electronic Product Reliability and Environmental Test Research Institute (CEPREI) (the fifth Electronics Research Institute of what was then the Ministry Information Industry of China) and CAPE (AVIC) both joined the Electronics Products and Systems Consortium of the university's CALCE in 2011.[66] Beihang University, formerly known as the Beijing University of Aeronautics and Astronautics (BUAA), was also a partner in CALCE's Prognostics and Health Management Group consortium.[67] CAPE remained a member of the CALCE consortium for three years, while BUAA ended its sponsorship sometime after joining. CALCE's website, updated in 2013, further does not list CEPREI as a member of its Electronic Products and Systems Consortium anymore.[68] However, CALCE maintains rela-

---

[64] "Embry-Riddle Aeronautical University, Civil Aviation University of China Sign Pact on Cooperative Education," Embry-Riddle Aeronautical University website, October 11, 2011.

[65] Janice Wood, "Embry-Riddle Partners with Chinese Universities," *General Aviation News*, November 30, 2011.

[66] "CEPREI, CAPE (AVIC) Join CALCE," CALCE (Center for Advanced Life Cycle Engineering), 2011.

[67] "CALCE PHM Consortium Members," CALCE Prognostics, undated.

[68] "CALCE EPS Consortium Members," CALCE (Center for Advanced Life Cycle Engineering), undated.

tions with AVIC and hosts occasional delegations from the latter.[69] Finally, the university runs a research park in collaboration with the Chinese Ministry of Science and Technology that focuses on health care, environment, agriculture, energy, and fire protection.[70]

The University of Michigan, ranked sixth among U.S. engineering schools, established a joint institute with Shanghai Jiao Tong University in 2006.[71] The institute has both undergraduate and graduate programs that focus on mechanical, electrical, and computer engineering.[72] It also has a number of centers and laboratories performing research in specific areas related to aerospace, such as a Dynamics and Vibration Laboratory and an Aero-Thermal Laboratory.[73]

Purdue University, ranked ninth among U.S. engineering schools, signed a collaboration agreement with China's Nanshan Group in June 2013. According to the founding document:

> The new agreement supports Purdue's Department of Aviation Technology and the Nanshan Aeronautical College to pursue educational programming, research and teaching projects through a proposed Purdue-Nanshan Institute for Global Aviation Studies.[74]

Additional information on this joint institute after 2013 is unavailable. In summer 2011, Purdue also signed an agreement with Beihang University to form the BUAA-Purdue Joint Laboratory on Energy Systems and the BUAA-Purdue Joint Laboratory on Low Emissions Com-

---

[69] Authors' correspondence with CALCE administrators

[70] "Overview," University of Maryland-China Research Park website, undated.

[71] "Joint Institute," Shanghai Jiao Tong University website, undated.

[72] "Undergraduate Program," University of Michigan-Shanghai Jiao Tong University Joint Institute, undated; and "Graduate Program Overview," University of Michigan-Shanghai Jiao Tong University Joint Institute website, undated.

[73] "Laboratories," University of Michigan-Shanghai Jiao Tong University Joint Institute website, undated.

[74] "Purdue Inks Partnerships with Nanshan, Tsinghua on Educational, Research Opportunities," Purdue University website, July 1, 2013.

bustion.[75] As of 2014, Purdue no longer has a formal relationship with Beihang as it is on the U.S. Department of Commerce's denied entities list related to rocket and missile technology concerns.[76]

### Other Connections

Stanford University, ranked second among U.S. engineering schools, opened a research center at Peking University in March 2012.[77] The center hosts a number of anchor programs, including the School of Engineering Internship and Travel Study programs. However, these programs do not appear to be connected to aerospace in any way.[78]

The University of California, Berkeley, ranked third among U.S. engineering schools, opened the Shanghai Zhangjiang Berkeley Engineering Innovation Center (Z-BEI) in cooperation with the Zhangjiang Hi-Tech Park in November 2013. Z-BEI aims to promote academic and industry research collaboration in such areas as information engineering, systems engineering, bioengineering, precision manufacturing, and green technology.[79] In October 2015, Berkeley and Tsinghua established the Tsinghua-Berkeley Shenzhen Institute, which houses research centers in three areas: environment and new energy, information technology and data science, and precision medicine and health care.[80]

Georgia Institute of Technology, ranked seventh among U.S. engineering schools, cooperated with Tianjin University to launch a

---

[75] Emily Venere, "Purdue, China Forming Joint Energy-Research Labs," Purdue University website, summer 2011.

[76] "Supplement No. 4 to Part 744—Entity List," Bureau of Industry and Security, September 7, 2016.

[77] Adam Gorlick, "Stanford Opens Research Center at Peking University," *Stanford Report*, March 22, 2012.

[78] "SCPKU Anchor Programs," Stanford Center at Peking University website, undated.

[79] Yu Wei, "UC-Berkeley Linked R&D Center Opens in Shanghai," *China Daily*, November 20, 2013.

[80] Rachel Cao Schafer, "Tsinghua-Berkeley Shenzhen Institute Inaugurated in China," Berkeley Engineering website, October 26, 2015.

master of science program in electrical and computer engineering at the Shenzhen Virtual University Park in April 2014.[81]

The University of Illinois at Urbana-Champaign, also ranked seventh among U.S. engineering schools, in April 2016 accepted an invitation from Zhejiang University to form a joint engineering institute. "Beginning in the fall 2016, about 150 undergraduate students will study civil engineering, computer engineering, electrical engineering, and mechanical engineering at the Institute."[82] In December 2013, the University of Illinois also "launched its first China office at the Shanghai Center. . . . Among other functions, the office will facilitate the admission process."[83] Finally, the University of Illinois signed a joint-education agreement with Zhejiang University in 2010 to establish a cooperative education program in agricultural and biological engineering and food sciences and human nutrition.[84]

The University of Southern California (USC), ranked 12th among U.S. engineering schools, "signed a Memorandum of Understanding (MOU) with the Science, Industry, Trade and Information Technology Commission of Shenzhen Municipality to explore educational and research collaborations in China" in October 2011.[85] Previously, in 2009, it signed the i-Podium Educational Cooperation Program with Peking University, which allows students from both universities to "take the same courses and interact via video streaming."[86] In 2008, it signed an MOU with Shanghai Jiao Tong University to begin establishing joint academic programs with educational exchanges mostly

---

[81] Chen Qide, "Georgia Tech Teams Up with Tianjin U," *China Daily*, April 29, 2014.

[82] "Engineering at Illinois and Zhejiang University Partner on Joint Institute for Engineering," Engineering at Illinois website, April 15, 2016.

[83] "US Universities Set Up Shop in China," *China Daily*, January 3, 2014.

[84] "U of I Signs Joint Education Agreement with Chinese University," Engineering at Illinois website, October 21, 2010.

[85] "USC Viterbi to Explore Research and Education Collaborations in Shenzhen, China," USC Viterbi School of Engineering website, November 1, 2011.

[86] "PKU and USC Sign i-Podium Educational Cooperation Program," Peking University, May 26, 2009.

focused on electrical engineering and communications technology.[87] USC also established an educational exchange program with Tsinghua University in June 2007. The program between USC's Viterbi School of Engineering and Tsinghua's School of Information Science and Technology offers a dual-degree program and facilitates student and faculty exchanges but does not appear related to aerospace.[88] Finally, USC Viterbi also has educational partnerships, established at unknown points in time, with both Beijing Jiaotong University and Xidian University, that appear to focus more on student exchanges.[89]

The University of California, Los Angeles, ranked 14th among U.S. engineering schools, established the Joint Research Institute (JRI) in Science and Engineering with Peking University in June 2009.[90] As of March 2012, collaborative projects between the two institutions through JRI have focused on biological and medical sciences, materials science and clean energy, information technology, environmental science, and earth and space science (such as astrophysics and cosmology).[91]

Virginia Polytechnic Institute and State University, ranked 21st among U.S. engineering schools, has a collaborative research lab with Shandong University. Established in September 2010, its research focuses on biophysics and bio-inspired technology.[92]

---

[87] "China's Shanghai Jiao Tong University and the Viterbi School Join Forces," USC Viterbi School of Engineering website, April 16, 2008.

[88] "Viterbi School Builds Strategic Bridges with Tsinghua, China's Top Technical University," USC Viterbi School of Engineering website, June 3, 2007; and "USC and Tsinghua University (THU) Program," USC Viterbi School of Engineering website, undated.

[89] "USC and Beijing Jiaotong University—Program," USC Viterbi School of Engineering website, undated; and "The Partnership," USC Viterbi School of Engineering website, undated.

[90] "About JRI," Joint Research Institute in Science and Engineering by Peking University and UCLA website, undated.

[91] "Collaborative Projects," Joint Research Institute in Science and Engineering by Peking University and UCLA website, undated.

[92] "Collaborative Research Lab Unveiled in China," Virginia Tech College of Engineering website, September 27, 2010.

The University of Washington, ranked 24th among U.S. engineering schools, set up a Global Innovation Exchange (GIX) with Tsinghua University in June 2015. Beginning in fall 2016, GIX will offer master's degrees in technology innovation. This partnership is also "the first time that a Chinese research university has established a physical presence in the United States."[93]

The Pennsylvania State University, ranked 27th among U.S. engineering schools, established the Joint Center for Energy Research with Dalian University of Technology in 2011.[94] The institute focuses on research on clean energy.

The Ohio State University, ranked 31st among U.S. engineering schools, established a gateway office in Shanghai in June 2010. The office appears to be responsible for managing the university's study-abroad program.[95]

The University of California, Davis, ranked 33rd among U.S. engineering schools, established the World Food Center–China in Zhuhai in cooperation with the city government in March 2015. The city of Zhuhai contributed $2.5 million to fund initial projects on improving food safety in China, a key concern for Chinese citizens.[96] In September 2014, it also established the ZEV (zero emission vehicles) Policy Lab with the China Automotive Technology and Research Center, which is overseen by SASAC.[97]

Boston University, ranked 35th among U.S. engineering schools, signed an agreement with Zhejiang University in October 2014. The agreement

---

[93] "UW and Tsinghua University Create Groundbreaking Partnership with Launch of the Global Innovation Exchange," *UW Today*, June 18, 2015.

[94] "Joint Center for Energy Research Promotes U.S.-China Collaboration," *Penn State News*, October 21, 2014.

[95] "News," Ohio State University website, undated.

[96] Pat Bailey, "Food Safety Agreement Sets Framework for World Food Center–China," World Food Center at UC Davis website, May 26, 2015.

[97] "China-U.S. ZEV Policy Lab," UC Davis China Center for Energy and Transportation website, October 23, 2015.

has developed a framework of cooperation between the College of Engineering at BU through its Center for Information and Systems Engineering (CISE) and the Faculty of Information Technology at Zhejiang through its State Key Laboratory of Industrial Control Technology.[98]

The University of California, Irvine (UCI), ranked 37th among U.S. engineering schools, in November 2015 "signed a Letter of Intent to develop a program of joint research and education efforts" with Dalian University of Technology, a highly ranked Chinese engineering university. UCI professors can teach classes at the Dalian joint institute and UCI students can take Dalian classes, while Dalian engineering students can finish their Dalian undergraduate degree at UCI and apply to UCI graduate programs.[99]

The University of Virginia, ranked 39th among U.S. engineering schools, established an office in 2013 to "strengthen its academic programs, research, internships, alumni engagement, and recruitment of students."[100]

The University of Florida, ranked 43rd among U.S. engineering schools, has 30 cooperative agreements with a range of Chinese institutions. The focus of these agreements ranges from research to exchanges to joint-degree programs. Their areas of research, meanwhile, range from agricultural life and sciences to emerging pathogens to sustainable infrastructure and environment. None concern aerospace, however.[101]

Arizona State University, ranked 46th among U.S. engineering schools, has partnerships with several Chinese universities, including Shandong University and Sichuan University. However, these do not concern aerospace. For example, it "established a joint Biodesign Center,

---

[98] Christina Polyzos, "Boston University and Zhejiang University Agreement," Boston University Center for Information and Systems Engineering website, November 5, 2014.

[99] "Samueli School to Collaborate with China's Dalian University of Technology," UCI Samueli School of Engineering website, December 4, 2015.

[100] "U.VA. in China: Officials to Celebrate Grand Opening of Shanghai Office," *UVA Today*, March 3, 2015.

[101] "UF Connections in China," University of Florida website, 2015.

in Qingdao, to focus on cancer and vaccine research, water and air purification systems and advanced explorations of nanotechnology."[102]

Case Western Reserve University, also ranked 46th among U.S. engineering schools, has student exchange partnerships with Hebei University of Technology and Tianjin University. The Case School of Engineering hosts students from Hebei University of Technology to complete their last two years for a joint undergraduate engineering degree and annually sponsors six to eight doctoral candidates, recipients of the Ministry of Education's China Scholarship Council, from Tianjin University to earn their doctorates.[103]

University of Notre Dame, ranked 48th among U.S. engineering schools, originally had plans to establish a joint liberal arts college with Zhejiang University, but these plans ultimately fell through in April 2016.[104]

Table 4.1 shows a list of the remaining top aerospace engineering programs that we investigated and found to have limited relationships with Chinese institutions.

### Chinese Universities' Relationships with U.S. Universities and Entities

We also looked at the top Chinese aerospace engineering programs for connections to U.S. entities. We started with the top ten programs; only three of the ten had connections, which were already accounted for in the survey of U.S. programs.

Among China's top ten aerospace engineering schools, only three have relationships with U.S. entities. The previous section provided details on these schools:

- Beihang University
- Zhejiang University
- Tsinghua University.

---

[102]"China Partnerships," Arizona State University website, undated.

[103]"Partnership Programs," Case School of Engineering website, undated.

[104]"University Abandons Plans to Establish Joint College in China," *Observer*, April 12, 2016.

The remaining seven do not appear to have relationships with U.S. entities. For reference, they are:

- School of Aeronautics in Northwestern Polytechnical University
- College of Aerospace Engineering in Nanjing University of Aeronautics and Astronautics
- School of Astronautics in Harbin Institute of Technology
- College of Aerospace Science and Engineering at the National University of Defense Technology
- School of Aerospace Engineering in Beijing Institute of Technology
- College of Aerospace and Civil Engineering in Harbin Engineering University
- School of Physics and Mechanical & Electrical Engineering in Xiamen University.

This reverse search for aviation connections between Chinese entities and U.S. universities produced no additional information on any relationships. It does suggest a dichotomy, where a few Chinese universities are striving to be global institutions connected with U.S. universities in general, although not necessarily out of concerns related to aviation or aerospace. Many other Chinese universities with leading aerospace programs are not connected to U.S. universities in any significant way.

## Findings on Chinese Investments and Relationships with U.S. Aviation

- Chinese investments in U.S. aviation have grown in scope and quantity over the past decade but are limited to smaller GA companies with technologies not particularly relevant to commercial or military aircraft, likely due to effective U.S. export and foreign-investment regulations.
- There are few special relationships between Chinese institutions and U.S. universities related to aviation beyond the normal pres-

ence of Chinese graduate students attending U.S. aerospace programs and existence of university-wide study-abroad and cultural exchanges.

In the next chapter, we review the findings of our conversations with subject-matter experts on the aviation industry, foreign investment in the United States, and on doing business in China. We discuss the implications for U.S. competitiveness and national security of these investments and relationships.

# CHAPTER FIVE

# Assessing the Effect of Chinese Investments in U.S. Aviation

This chapter assesses the economic, technological, and military effects of Chinese investments in the United States. Our assessment is informed by consultations with subject-matter experts and a careful reading of the open-source literature. We spoke with nine subject-matter experts with expertise in the aviation and aerospace industries, U.S.-China business relations, U.S. trade restrictions and export controls, and legal regulations on foreign investment in U.S. companies. We solicited their insights on four broad aspects of Chinese investments in U.S. aviation: (1) whether China had a strategy, (2) the potential for technology transfer accelerating the development of China's domestic aviation/aerospace industry, (3) the effect on U.S. global competitiveness, and (4) the implications for Chinese military capabilities. One of the authors also attended the U.S.-China Aviation Summit in June 2016, listening to panels and talking offline with attendees to further supplement our findings.[1]

In addition to offering insights into Chinese investment in U.S. aviation, the subject-matter experts also provided a global view of Chinese investment patterns, particularly in European aviation and aerospace firms. While these went beyond the scope of this study, they allowed the team to draw nuances of the implications of Chinese investments across different platforms and subsectors (e.g., engines, avionics), as well as across countries and regions.

---

[1] Seventh U.S.-China Aviation Summit, Washington, D.C., June 19–21, 2016.

Some of the experts we spoke with and agreed to be named include:

- Richard Aboulafia, vice president of analysis, Teal Group
- Ronald Epstein, senior equity analyst, Bank of America Merrill Lynch
- Sash Tusa, defense analyst, Agency Partners, United Kingdom
- Joe Borich, senior adviser, Nyhus
- Richard Bitzinger, senior fellow and coordinator of the Military Transformations Program, S. Rajaratnam School of International Studies, Nanyang Technological University, Singapore.

Given the GA nature of most of the investments by Chinese firms to date, there are few technology-transfer concerns. The main benefits to China's industry would be on the business-process side, such as international marketing, achieving FAA safety certifications, and product support. U.S. competitiveness is unlikely to be threatened in the near term because production of China's LCA—the C919—may be further delayed and operate less efficiently than current narrow-body aircraft on the international market. However, some experts remain concerned about the transfer of engine or avionics technology through COMAC C919 joint ventures with Western companies; others think technology transfers are unlikely given U.S. export controls. A more-competitive civil aviation industry broadly supports Chinese military aviation, for example, with a larger talent pool, scales of efficiency, and greater supply chain options. However, direct military implications are minimal because advanced commercial-aviation technology differs from military-aviation technologies (e.g., stealth, radar, supersonic engines).

## A Strategy for Chinese Investments in U.S. Aviation?

The subject-matter experts we spoke with observed that the identified acquisitions and investments were modest, especially when compared with U.S. deals with other countries (e.g., Japan). Indeed, the subject-matter experts were not immediately familiar with many of the U.S.-Chinese mergers, acquisitions, and joint ventures RAND presented in its research. Upon review, experts described the Chinese strategy—if there was one—as an opportunistic "take what you can get" approach. The subject-matter experts with aviation expertise noted that most transactions involved GA companies rather than commercial aircraft manufacturers or their supply chains. They noted that GA technologies, such as piston engines or fiberglass structures, are not particularly relevant to commercial or military aircraft.

The limited market for and utility of investing in small GA aircraft prompts questions as to whether these investments are driven by political or business logic. While the government has signaled a clear interest in expanding GA in China, it remains a very limited market within China and relatively small compared with commercial aviation globally. According to experts, there are a variety of obstacles to China's developing a GA culture like the United States: the military and government exert strong control over the airspace; limited infrastructure makes operating a GA aircraft in China cost twice as much as it would in the United States;[2] and a recent crackdown on corruption has further depressed the number of people who would be willing and able to make such high-profile, expensive purchases.[3] These factors all result in a very limited market for business jets, turboprops, and smaller propeller aircraft.

The subject-matter experts with backgrounds on the context of Asia and China overwhelmingly disclaimed the idea of any robust, centrally coordinated approach to Chinese firms' investments in the U.S. aviation and aerospace industries. One subject-matter expert expressed skepticism that there was any "concerted, top-down, guided effort to deliberately cherry-pick the best technologies." When discussing China's strategies and goals in aviation, these subject-matter experts down-

---

[2] "Breakout IV: China and US Partnership Opportunities for Aviation Development," question-and-answer session during panel, Seventh U.S.-China Aviation Summit, Washington, D.C., June 19–21, 2016.

[3] Sideline conversation, Seventh U.S.-China Aviation Summit, Washington, D.C., June 19–21, 2016.

played Chinese investments in U.S. GA firms and focused on joint ventures related to COMAC's C919 program (as shown in Chapter Two, Table 2.1).

## Technology Transfer and Development of China's Domestic Industry

In general, these Chinese mergers, acquisitions, and joint ventures are not considered sources of highly innovative or complex aerospace products or components. The majority of these investments—and we explore notable exceptions later in this discussion—pertain to low-tech, older technology for smaller GA platforms.

Subject-matter experts expressed a range of opinions about whether or not U.S.-Chinese joint ventures could unintentionally lead to technology transfers beneficial to China's domestic aviation industry. Specifically, some subject-matter experts were concerned about joint ventures in the areas of engines and avionics. Those subject-matter experts argued that because of their indefinite nature, joint ventures can provide Chinese firms with more access to technology, research, and other benefits either intentionally or unintentionally versus a simple integrator-supplier relationship. Other subject-matter experts, however, countered that strict U.S. export controls, along with intellectual property concerns and interests, limit the prospect of sensitive technology transfers, even in joint ventures.

There are at least three ways by which China can acquire more technology in aviation or any other sector: build it, buy it, or steal it. In other words, China can figure out how to build better, more technologically sophisticated aircraft indigenously, acquire this technology via foreign purchases, or steal it through industrial espionage. Historically, indigenous means have had demonstrated significant limitations for the PRC, and overseas purchases of military aviation have been circumscribed because of export restrictions imposed by the United States and other Western governments. Beyond figuring out ways to circumvent these export controls, industrial espionage has remained a high Chinese priority.[4]

The COMAC C919 and ARJ21 projects have a large number of U.S. suppliers and U.S.-China joint ventures.[5] Some subject-matter experts expressed concerns that joint ventures with U.S. companies for the C919's avionics systems and engines could result in technology transfers. One subject-matter expert described the C919 project as a key route for China to acquire Western manufacturing technology, despite export controls and restrictions on technology transfers.

Some subject-matter experts also claimed that mergers and acquisitions, but particularly joint ventures, could offer Chinese firms an opportunity to gain management insights from Western businesses. Improvements in domestic Chinese firms' business practices, management approaches, and manufacturing processes could increase efficiencies and global competitiveness.

## Implications for U.S. Global Competitiveness

U.S. aviation jobs and manufacturing capacity depend on remaining globally competitive. Most subject-matter experts we interviewed do not see Chinese firms posing a serious threat to U.S. or Western global competitiveness in the near term (five to ten years). In the long term, as Chinese firms continue to develop new platforms, some subject-matter experts do see prospects for increased Chinese global competitiveness.

While China has a very small market for GA, Chinese demand currently provides about one-fifth of the global market for LCA. As

---

[4]  Phillip C. Saunders and Joshua Wiseman, "China's Quest for Advanced Aviation Technologies," in Richard P. Hallion, Roger Cliff, and Phillip C. Saunders, eds., *The Chinese Air Force: Evolving Concepts, Roles, and Capabilities*, Washington, D.C.: National Defense University Press, 2012, pp. 271–323. See also William C. Hannas, James Mulvenon, and Anna B. Puglisi, *Chinese Industrial Espionage: Technological Acquisition and Military Modernization*, New York: Routledge, 2013.

[5]  ARJ21: 15 of 22 suppliers are from United States; C-919: 29 of 40 suppliers are from United States (from Seventh U.S.-China Aviation Summit, Washington, D.C., June 19–21, 2016).

COMAC develops more commercial aircraft, a medium- to long-term shift away from Western suppliers could have a negative effect on U.S. global market share. Experts posit that even 50 percent of the Chinese market's shifting to domestically produced jets would have significant repercussions for Boeing and Airbus.

One expert, however, also noted that it is unlikely that C919 production can be sustained without ongoing government subsidies.[6] Aerospace manufacturing requires highly skilled labor because of advanced manufacturing processes, which makes the Chinese competitive advantage of cheap labor less relevant. The introduction of robotics or automation does not significantly change that for aviation, given low production rates and the fact that automation requires even more high-skilled labor to sustain. Designing and building aircraft for international markets requires high attention to detail and quality-control standards to gain FAA/EASA certification. Cost-saving measures, should they compromise quality and reliability, would be unacceptable. Furthermore, as China's labor costs and wages increase, any competitive advantage from cheap labor or a low cost of living continues to erode.

In the U.S. market, subject-matter experts did not predict large-scale Chinese investment in the GA or LCA sectors. Some experts did point to the potential marketability of small Chinese GA or utility aircraft in the United States. The latest version of the Harbin Y-12, a twin-engine turboprop produced by an AVIC subsidiary, received FAA certification, and a California-based tourism company reportedly ordered 20 of the planes two years ago. However, there is no evidence that any have been delivered or that any are being operated commercially in the United States.

Overall, while experts do not anticipate Chinese aviation and aerospace industries eating into U.S. global competitiveness, jobs, or manufacturing in the near term, should a mid- to long-term shift in

---

[6] While there is debate over Western government financial support to Boeing and Airbus that has been the subject of multiple World Trade Organization suits, those funds generally support basic aviation research and aircraft development or are government guarantees that reduce the cost of capital, not production costs that exceed revenue.

Chinese domestic market share occur, it could enable a challenge to U.S. global competitiveness in commercial aviation.

## Possible Military Implications

U.S. export controls and regulations seek to mitigate risks to U.S. national security that may emanate from foreign investments. In general, experts expressed confidence in the oversight and restrictions overseen by the U.S. Department of State's International Traffic in Arms Regulation (ITAR) and CFIUS. In fact, the concern that a potential deal may come under CFIUS scrutiny has been enough to scuttle some deals, according to multiple experts we interviewed.

In most cases, subject-matter experts assessed that the technology China has been acquiring through these investments would have quite limited—if any—military applications. Experts did note, however, that there is some potential for China to leverage the investments, especially joint ventures, to advance indigenous unmanned aerial vehicle technology, domestic engine development, and avionics capabilities. Given China's preference for indigenous production, foreign suppliers, shown in Table 2.1 (Chapter Two) and Table 5.1, suggest shortfalls in Chinese aviation manufacturing capabilities. Additionally, investments in small U.S. companies producing light helicopters (e.g., Brantly and Enstrom) and small planes could improve Chinese military scout helicopters and small maritime reconnaissance aircraft, for example. Such technologies are broadly available, and blocking their acquisition would have led Chinese companies to simply procure from other foreign providers.

One area where experts expressed potential concerns was over Chinese acquisition of Western engine and avionics technology through joint ventures. GE Avionics, through a joint venture, provides the common core avionics system for the C919 in development. GE also provides the common core system for the Boeing 787. Two experts noted that the C919's avionics core processing system was initially developed by Smiths Aerospace, a UK company, which was acquired by GE, for the Eurofighter Typhoon and subsequently applied to the F-22. While there is no reason to believe the military technology has

**Table 5.1**
**C919 Partners Grouped by Technology**

| | |
|---|---|
| Avionics C919 partners | GE-AVIC joint venture, Hamilton Sunstrand, Honeywell International |
| Engine C919 partners | CFM International, which is joint between GE, Sanfran (France), and Thales (France) |

SOURCE: Based on Table 2.1.

been transferred, that modern commercial-aviation avionics based on technology was also applied to modern jet fighters demonstrates that there is some overlap between civil- and military-aviation technology. One expert expressed concern that Chinese joint ventures could enable China to reduce its disadvantages in the areas of engine performance, operational longevity, and ease of maintenance.

There is always a possibility for some civil-military integration or comingling of civilian technologies with the military side. This is a very real concern in China because "civil-military integration" continues to be emphasized by PRC political and military leaders.[7] Although severe stovepiping characterizes all aspects of Chinese bureaucratic systems, civil-military linkages are real and numerous. In the aviation sector, supply chains and key components are sometimes identical or overlapping.[8] Moreover, funding, ownership, and control of Chinese companies are often very difficult to discern. While COMAC was established as a purveyor of civilian aircraft, the Chinese military has influence through AVIC's investment in the C919 project as well as through broader government ownership of COMAC.[9] Furthermore, a modified configuration of an aircraft such as the C919 could be used as a

---

[7] See, for example, the discussion and analysis in Tai Ming Cheung, *Fortifying China: The Struggle to Build a Modern Defense Economy*, Ithaca, N.Y.: Cornell University Press, 2009; and Tai Ming Cheung, *Forging China's Military Might: A New Framework for Assessing Innovation*, Baltimore, Md.: Johns Hopkins University Press, 2014.

[8] Shen Pin-Luen, "China's Aviation Industry: Past, Present and Future," in Richard P. Hallion, Roger Cliff, and Phillip C. Saunders, eds., *The Chinese Air Force: Evolving Concepts, Roles, and Capabilities*, Washington, D.C.: National Defense University Press, 2012, p. 260.

[9] Cheung, 2009, p. 123.

military transport—and strategic airlift constitutes a severe deficiency for the People's Liberation Army.[10]

The business and manufacturing best practices gleaned from and exploited in the commercial and GA markets may also benefit the military-industrial base; streamlining processes and implementing best practices may lead to more efficient and effective military systems. Additionally, where products are obviously dual use, this bleed-over will likely occur (with scout helicopters, for example). However, in the Chinese case, experts expressed skepticism that the Chinese military would gain from the vast majority of these investments because parallel military programs require products, systems, and components of a significantly higher degree of capability.

In general, experts expressed confidence in the ability of ITAR and CFIUS restrictions to flag and prevent technology transfers of concern to the United States, even in the case of joint ventures. Industry and company representatives stressed how attentive they are to technology-transfer concerns throughout the life cycle of a joint venture not only for legal reasons but to protect intellectual property.

## Findings on the Implications of Chinese Investments

- On the whole, aviation- and aerospace-technology experts are not concerned about Chinese investment in U.S. aviation's effect on U.S. global competitiveness or the potential for technology transfers that might have military implications.
- The majority of experts do not see a grand strategy, directed by the central government, for Chinese firms to pursue investments in U.S. aviation and aerospace interests. The Chinese-U.S. aviation mergers and acquisitions involve small companies producing relatively low-tech products.

---

[10] Michael S. Chase, Jeffrey Engstrom, Tai Ming Cheung, Kristen Gunness, Scott Warren Harold, Susan Puska, and Samuel K. Berkowitz, *China's Incomplete Military Transformation: Assessing the Weaknesses of the People's Liberation Army*, Santa Monica, Calif.: RAND Corporation, RR-893-USCC, 2015, p. 113; and Cheung, 2009, p. 123.

- Joint ventures generated more debate among experts. Some experts contend joint ventures in the areas of engines and avionics could result in technology transfers—some with potential military dimensions. Other experts insist that CFIUS, ITAR, and intellectual-property considerations incentivize U.S. participants to maintain strict technology controls when participating in joint ventures.[11]
- Experts do not see Chinese aviation and aerospace firms encroaching on U.S. global competitiveness in the near term (five to ten years). Should a longer-term shift in domestic Chinese LCA market share occur, with Chinese-made airframes displacing purchase of U.S.-built aircraft, it could have a serious negative effect on U.S. firms.

## Policy Implications

While this study has not found any immediate concerns with Chinese investment in U.S. aviation to date, given China's aggressive aviation industrial policies, it is prudent to continue monitoring the effectiveness of U.S. export controls and foreign investment regulation. Concerns remain about technology transfer through covert operations or Chinese investment in non-U.S. companies in the aviation global supply chain. With no additional recommendations specific to Chinese investment in U.S. aviation, we want to highlight previous recommendations from past RAND studies.

While there is no immediate expectation that China will achieve its goal of being a viable global competitor in the commercial aviation market, U.S. policymakers can take several steps to minimize the

---

[11] Finding conclusive evidence that legal restrictions are preventing Chinese appropriation of U.S. technology is difficult as applications for CFIUS and export approvals are proprietary. Even more so, many potential investments are not even pursued because of perceived risks of seeking those approvals.

potentially distorting effects of PRC industrial policies. Such steps rec-
ommended in previous RAND studies include the following:[12]

- Engage the European Union to establish a consensus on aerospace industrial policy norms. There are unresolved disagree-
ments about government support to Boeing and Airbus. Without
consensus, it is hard to hold China accountable to any standards.
- Work toward improving transparency of Chinese aerospace actors
by providing more clarity on aircraft purchases by Chinese state-
owned airlines, implementing more intellectual-property safe-
guards in the context of component certifications by the FAA or
EASA, and increasing voluntary reporting by U.S. suppliers that
have China-based operations on how investment decisions have
been influenced by PRC industrial policy.
- Continue to monitor PRC aerospace industrial policy and work
through bilateral and World Trade Organization forums to elimi-
nate, in general, industry-specific policies and, in particular, to
prevent these industrial policies from supporting the entry of the
C919 or future COMAC aircraft into foreign markets.

U.S. government policies also play a role in supporting U.S. aviation industries broadly, not specific to Chinese competition. Assessing
the optimal nature of that is beyond the scope of this report, but recent
policy uncertainty does not help. For example:

- the recent lapse in the Import/Export Bank authorization from
U.S. Congress, which provided significant support to commercial
aviation
- the instability of funding to U.S. aviation R&D, where NASA
aeronautics funding is less than one-third of 1990s levels. The
fiscal year 2017 President's Budget significantly increases future

---

[12] Crane et al., 2014.

aeronautics funding, but without a congressional budget those increases are more uncertain than ever.[13]

---

[13] Philip S. Anton, Liisa Ecola, James G. Kallimani, Thomas Light, Chad J. R. Ohlandt, Jan Osburg, Raj Raman, and Clifford A. Grammich, *Advancing Aeronautics: A Decision Framework for Selecting Research Agendas*, Santa Monica, Calif.: RAND Corporation, MG-997-NASA, 2011; and Jan Osburg, Philip S. Anton, Frank Camm, Jeremy M. Eckhause, Jaime Hastings, Jakub Hlavka, James G. Kallimani, Thomas Light, Chad J. R. Ohlandt, Douglas Shontz, Abbie Tingstad, and Jia Xu, *Expanding Flight Research Capabilities, Needs, and Management Options for NASA's Aeronautics Research Mission Directorate*, Santa Monica, Calif.: RAND Corporation, RR-1361-NASA, 2016.

# CHAPTER SIX

# Conclusions

Over the past decade, Chinese investors have ventured outside of China by acquiring U.S. aviation companies, previously unheard of given that most investment was into China. Although CFIUS or export controls appeared to have been followed in all cases, these investments raise concerns of inadvertent technology transfer that might undermine U.S. national security and competitiveness. The publicly identified investments are limited to GA manufacturers with less advanced technologies and do not pose competitiveness challenges or national-security concerns. Chinese institutional links with U.S. universities related to aviation were found to be limited. However, China continues to operate with an unambiguous government policy toward advancing its global competitiveness in aviation and to focus considerable resources towards the production of LCA. Concerns about U.S. competitiveness should remain centered around C919-related joint ventures or future Chinese LCA designs, such as wide-body aircraft development with the Russians. Concerns about U.S. national-security issues should remain focused on espionage, cybercrime, and illegal technology transfers.

After assessing Chinese future demand for aviation products and the state of Chinese domestic aviation production, documenting Chinese aviation industrial policy, reviewing recent Chinese investments in U.S. aviation and U.S. aviation–related university connections with Chinese entities, and discussing the implications of those investment with experts, our main findings are as follows:

- China will likely account for up to one-fifth of global demand for LCA and is trying to grow its domestic GA industry, which is currently underdeveloped.
- China has unambiguous policy driving a whole-of-government effort to develop a globally competitive aviation industry by producing LCA and expanding China's domestic GA market.
- Chinese investments in U.S. aviation have grown in scope and quantity over the past decade but are limited to smaller GA companies with technologies not particularly relevant to commercial or military aircraft, likely because of effective U.S. export and foreign-investment regulations.
- There are few special relationships between Chinese institutions and U.S. universities related to aviation beyond the normal presence of Chinese graduate students attending U.S. aerospace programs and existence of university-wide study-abroad and cultural exchanges.
- Given the GA nature of most of the investments by Chinese firms to date, there are few technology-transfer concerns. The main benefits to China's industry would be on the business-process side, such as international marketing, achieving FAA safety certifications, and product support.
- U.S. competitiveness is unlikely to be threatened in the near term because production of China's LCA—the C919—may be further delayed and operate less efficiently than current Western narrow-body aircraft on the international market. However, some experts remain concerned about the transfer of engine or avionics technology through COMAC C919 joint ventures with Western companies; others think technology transfers are unlikely given U.S. export controls.
- A more competitive civil aviation industry broadly supports Chinese military aviation (e.g., larger talent pool, scales of efficiency, greater supply chain options). However, direct military implications are minimal because advanced commercial-aviation technology differs from military-aviation technologies (e.g., stealth, radar, supersonic engines).

# References

"12th Five-Year Plan for Chinese Civil Aviation Development [中国民用航空发展第十二个五年规划 [2011-2015年]]," Civil Aviation Administration of China, April 2, 2011.

ABCDlist, "COMAC C919 Production List," ABCDlist website, June 7, 2016.

"About JRI," Joint Research Institute in Science and Engineering by Peking University and UCLA website, undated. As of July 1, 2016:
http://www.pku-jri.ucla.edu/jri/about

"About Us [关于我们]," COMAC website, undated. As of June 12, 2016:
http://www.comac.cc/gywm/gsjj/

Aerospace Industries Association, *U.S. Aerospace Trade Balance*, 2016. As of June 24, 2016:
http://www.aia-aerospace.org/assets/Series_08_-_Balance.pdf

Airbus, "Airbus Results 2015," website, 2016. As of June 29, 2016:
http://www.airbus.com/presscentre/corporate-information/key-documents

———, "Global Market Forecast: Flying by Numbers 2015–2034," website, 2016. As of June 29, 2016:
http://www.airbus.com/company/market/forecast/

Airframer, "COMAC C919," website, May 5, 2016. As of July 15, 2016:
http://www.airframer.com/aircraft_detail.html?model=C919

Anton, Philip S., Liisa Ecola, James G. Kallimani, Thomas Light, Chad J. R. Ohlandt, Jan Osburg, Raj Raman, and Clifford A. Grammich, *Advancing Aeronautics: A Decision Framework for Selecting Research Agendas*, Santa Monica, Calif.: RAND Corporation, MG-997-NASA, 2011. As of December 7, 2016:
http://www.rand.org/pubs/monographs/MG997.html

"AVIC International Expands Commercial Aerospace Services Portfolio with the Acquisition of Align Aerospace," AVIC International website, March 31, 2015. As of July 1, 2016:
http://www.avic-intl.cn/detail.aspx?cid=2245

Bailey, Pat, "Food Safety Agreement Sets Framework for World Food Center–China," World Food Center at UC Davis website, May 26, 2015. As of July 1, 2016:
http://worldfoodcenter.ucdavis.edu/news/world_food_center_china.html

"Best Bush Planes: Flying Cessna, Piper, Beech, DeHavilland, Airplanes and Aircraft," Bush-planes.com, undated. As of August 30, 2016:
http://www.bush-planes.com/

"BHR and AVIC Auto Acquire Henniges Automotive," PR Newswire, September 15, 2015. As of August 29, 2016:
http://www.prnewswire.com/news-releases/bhr-and-avic-auto-acquire-henniges-automotive-300143072.html

Bloomberg, "China Reform Holdings Corp Ltd," company, undated. As of June 8, 2016:
http://www.bloomberg.com/profiles/companies/1002186D:CH-china-reform-holdings-corp-ltd

Boeing, Boeing commercial, homepage, undated. As of June 29, 2016:
http://www.boeing.com/commercial

———, "Current Market Outlook 2015–2034," website, 2015. As of June 29, 2016:
http://www.boeing.com/resources/boeingdotcom/commercial/about-our-market/assets/downloads/Boeing_Current_Market_Outlook_2015.pdf

Bombardier, "Commercial Aircraft Status Reports," website, undated. As of September 16, 2016:
http://www.bombardier.com/en/media/commercial-aircraft-status-reports.html

Brantly B-2B Helicopter, "Service Bulletin 111," website, February 28, 2011. As of July 1, 2016:
http://www.brantly.com/index.html

"Breakout IV: China and US Partnership Opportunities for Aviation Development," question-and-answer session during panel, Seventh U.S.-China Aviation Summit, Washington, D.C., June 19–21, 2016.

Bureau of Economic Analysis, Department of Commerce, "U.S. Trade in Goods and Services, 1992–Present," July 6, 2016. As of July 18, 2016:
http://www.bea.gov/newsreleases/international/trade/trad_time_series.xls

"CAAC Minister Feng Zheng Lin Visits COMAC Lab: Fully Supports C919 Research [中国民航局局长冯正霖到中国商飞调研：全力支持C919研制工作]," COMAC News Center, May 10, 2016. As of June 11, 2016:
http://www.comac.cc/xwzx/gsxw/201605/10/t20160510_3805757.shtml

"CALCE EPS Consortium Members," CALCE (Center for Advanced Life Cycle Engineering), undated. As of July 1, 2016:
http://www.calce.umd.edu/general/membership/members.html

"CALCE PHM Consortium Members," CALCE Prognostics, undated. As of July 1, 2016: http://www.prognostics.umd.edu/memberslist.htm

Cantle, Katie, "COMAC Establishes Finance Lease Company for C919 Sales," *China Aviation Daily*, May 16, 2012. As of June 10, 2016: http://www.chinaaviationdaily.com/news/19/19281.html

———, "AVIC, GE Aviation formally launch integrated avionics JV," *Air Transport World*, October 22, 2012. As of December 14, 2016: http://atwonline.com/operations/ avic-ge-aviation-formally-launch-integrated-avionics-jv

"CEPREI, CAPE (AVIC) Join CALCE," CALCE (Center for Advanced Life Cycle Engineering), 2011. As of July 1, 2016: http://www.calce.umd.edu/whats_new/2011/CAPE_CEPREI_CALCE_join.html

"Cessna and CAIGA Joint Venture to Start Operations [赛斯纳和中航通用飞机的合资企业即将运营]," *China Daily* [中国日报], April 16, 2013. As of July 1, 2016: http://auto.chinadaily.com.cn/planeyacht/2013-04/16/content_16411869.htm

Chang, Iris, *Thread of the Silkworm*, New York: Basic Books, 1995.

Chase, Michael S., Jeffrey Engstrom, Tai Ming Cheung, Kristen Gunness, Scott Warren Harold, Susan Puska, and Samuel K. Berkowitz, *China's Incomplete Military Transformation: Assessing the Weaknesses of the People's Liberation Army (PLA)*, Santa Monica, Calif.: RAND Corporation, RR-893-USCC, 2015. As of July 19, 2016: http://www.rand.org/pubs/research_reports/RR893.html

Chen Qide, "Georgia Tech Teams Up with Tianjin U," *China Daily*, April 29, 2014. As of July 1, 2016: http://europe.chinadaily.com.cn/business/2014-04/29/content_17472204.htm

Cheung, Tai Ming, *Fortifying China: The Struggle to Build a Modern Defense Economy*, Ithaca, N.Y.: Cornell University Press, 2009.

———, *Forging China's Military Might: A New Framework for Assessing Innovation*, Baltimore, Md.: Johns Hopkins University Press, 2014.

"China Development Bank Financial Leasing Co., Ltd," application proof, *Hong Kong Exchange News*, February 2016. As of June 10, 2016: http://www.hkexnews.hk/app/SEHK/2016/2016022602/Documents/ SEHK201602260017.pdf

China Exim Bank, "Export-Import Bank of China," website, undated. As of January 22, 2017: http://english.eximbank.gov.cn/tm/en-TCN/index_617.html

"China: Homegrown C919 Jets' Final Assembly Line Settles in Shanghai's Pudong" Xinhua, November 19, 2009.

"China Partnerships," Arizona State University website, undated. As of July 1, 2016:
https://global.asu.edu/china-partnerships

"China Reform Holdings Corporation Liu Dongsheng Visits COMAC [中国国新董事长刘东生到中国商飞访问]," COMAC News Center, May 27, 2016. As of June 12, 2016:
http://www.comac.cc/xwzx/gsxw/201605/27/t20160527_3878449.shtml

"China Succeeds in Its Target Unmanned Helicopter's First Flight," *Global Times*, May 8, 2011. As of December 5, 2016:
http://www.globaltimes.cn/content/652696.shtml

"China Unveils Jetliner in Bid to Compete with Boeing, Airbus," Associated Press, November 2, 2015.

"China-U.S. ZEV Policy Lab," UC Davis China Center for Energy and Transportation, October 23, 2015. As of July 1, 2016:
http://chinacenter.ucdavis.edu/initiatives/china-u-s-zev-policy-lab

"China Wins 100 C919 Orders, Breaks Airbus-Boeing Grip," *Bloomberg News*, November 16, 2010. As of September 7, 2016:
http://www.bloomberg.com/news/articles/2010-11-16/china-planemaker-gets-100-c919-aircraft-orders-from-ge-lessors-airlines

"China's Shanghai Jiao Tong University and the Viterbi School Join Forces," USC Viterbi School of Engineering website, April 16, 2008. As of July 1, 2016:
http://viterbi.usc.edu/news/news/2008/china-s-shanghai.htm

"Chinese Firm Completes U.S. Aircraft Maker Merger," *China Daily*, October 17, 2013. As of July 1, 2016:
http://usa.chinadaily.com.cn/business/2013-10/17/content_17040096.htm

Chongqing Helicopter Manufacturing Investment Company, "CQHIC Acquires U.S. Enstrom [重庆直投收购美国恩斯特龙]," website, January 5, 2013. As of July 1, 2016:
http://www.cqhic.cn/aspx/html/show.aspx?classid=11&id=313

Cliff, Roger, Chad J. R. Ohlandt, and David Yang, *Ready for Takeoff: China's Advancing Aerospace Industry*, Santa Monica, Calif.: RAND Corporation, MG-1100-UCESRC, 2011. As of June 29, 2016:
http://www.rand.org/t/mg1100

Clover, Charles, "UK Universities Under Scrutiny over China Ties," *Financial Times*, June 23, 2015. As of July 1, 2016:
https://www.ft.com/content/af5ea60e-1578-11e5-be54-00144feabdc0

"Collaborative Projects," Joint Research Institute in Science and Engineering by Peking University and UCLA website, undated. As of July 1, 2016:
http://www.pku-jri.ucla.edu/jri/article/124994

"Collaborative Research Lab Unveiled in China," Virginia Tech College of Engineering website, September 27, 2010. As of July 1, 2016:
https://www.eng.vt.edu/news/collaborative-research-lab-unveiled-china

COMAC Shanghai Aviation Flight Test Center News Center, homepage, undated. As of June 6, 2016:
http://www.comac.cc/xw/jcxx/sfzx/

"COMAC Signs Financing Framework Agreement with TEIBC," COMAC news website, October 10, 2015. As of January 22, 2017:
http://english.comac.cc/news/latest/201510/10/t20151010_2930701.shtml

"COMAC Signs Strategic Cooperation Agreement with SPD Bank," COMAC News, January 28, 2016. As of June 11, 2016:
http://english.comac.cc/news/latest/201602/01/t20160201_3437945.shtml

"Company Profile: Hanxing Group," undated. As of December 5, 2016:
http://glasairaviation.com/wp-content/uploads/2016/08/Hanxing_Group_Profile.pdf

Continental Motors, "Continental Motors Services Acquires United Turbine and UT Aeroparts Corporations," February 2, 2015.

"Continental Motors Services Acquires United Turbine and UT Aeroparts Corporations," Continental Motors, February 2, 2015.

Crane, Keith, Jill E. Luoto, Scott Warren Harold, David Yang, Samuel K. Berkowitz, and Xiao Wang, *The Effectiveness of China's Industrial Policies in Commercial Aviation Manufacturing*, Santa Monica, Calif.: RAND Corporation, RR-245, 2014. As of July 18, 2016:
http://www.rand.org/pubs/research_reports/RR245.html

De La Merced, Michael J., "A.I.G. Sells Aircraft Leasing Unit for $5.4 Billion," *New York Times*, December 15, 2013. As of July 1, 2016:
http://dealbook.nytimes.com/2013/12/15/a-i-g-said-near-deal-to-sell-aircraft-leasing-unit/?_r=0

"Director of Sichuan SASAC Xu Jin Visits COMAC [四川省国资委主任徐进到中国商飞访]," COMAC News Center, June 6, 2016. As of June 9, 2016:
http://www.comac.cc/xwzx/gsxw/201606/06/t20160606_3912475.shtml

"Education and Advice," *U.S. News and World Report*, undated. As of December 12, 2016:
http://www.usnews.com/education

Embraer, "Embraer Releases Fourth Quarter and Fiscal Year 2015 Results and 2016 Outlook," São Paulo, March 3, 2016. As of September 16, 2016:
http://www.embraer.com.br/Documents/noticias/Release%20US%204Q15_FINAL.pdf

"Embry-Riddle Aeronautical University, Civil Aviation University of China Sign Pact on Cooperative Education," Embry-Riddle Aeronautical University website, October 11, 2011. As of July 1, 2016:
http://dts.erau.edu/newsroom/press-releases/embry-riddle-aeronautical-university-civil-aviation-university-of-china-sign-pact-on-cooperative-education.html

"Engineering at Illinois and Zhejiang University Partner on Joint Institute for Engineering," Engineering at Illinois website, April 15, 2016. As of July 1, 2016:
http://engineering.illinois.edu/news/article/16405

"Exhibition: 'Qian Xue-Sen: A Man of Science and an Inspiration to Scholars,'" Caltech website, undated. As of July 1, 2016:
https://www.caltech.edu/content/
exhibition-qian-xue-sen-man-science-and-inspiration-scholars

Fallows, James, *China Airborne*, New York, N.Y.: Pantheon Books, 2012.

Federal Aviation Administration, "General Aviation Airports: A National Asset," website, May 2012. As of July 5, 2016:
http://www.faa.gov/airports/planning_capacity/ga_study/

Flottau, Jens, Michael Bruno, Graham Warwick, Guy Norris, and Bradley Perrett, "Subsidy Battle Anew," *Aviation Week and Space Technology*, July 4–17, 2016.

"GE and AVIC Joint Venture Creates New Global Business Opportunities," GE Aviation website, November 16, 2009. As of July 1, 2016:
http://www.geaviation.com/press/other/other_20091116.html

"GE and AVIC Sign Agreement for Integrated Avionics Joint Venture," GE Aviation website, January 21, 2011. As of July 1, 2016:
http://www.geaviation.com/press/systems/systems_20110121.html

Ge, Lena, "China-Made Y-12F Turboprop Aircraft Gets FAA Type Certification," *China Aviation Daily*, February 25, 2016. As of July 5, 2016:
http://www.chinaaviationdaily.com/news/51/51061.html

"General Office of Inspection and Supervision Team Inspects COMAC [国办督查组到中国商飞调研]," COMAC News Center, May 30, 2016. As of June 9, 2016:
http://www.comac.cc/xwzx/gsxw/201605/30/t20160530_3886049.shtml

Gorlick, Adam, "Stanford Opens Research Center at Peking University," *Stanford Report*, March 22, 2012. As of July 1, 2016:
http://news.stanford.edu/news/2012/march/peking-university-center-032212.html

Govindasamy, Siva, and Matthew Miller, "Exclusive: China-Made Regional Jet Set for Delivery, but No U.S. Certification," Reuters, October 21, 2015.

Goyer, Robert, "New Owners for Cirrus," *Flying*, February 28, 2011. As of July 1, 2016:
http://www.flyingmag.com/news/new-owners-cirrus

"Graduate Program Overview," University of Michigan-Shanghai Jiao Tong University Joint Institute website, undated. As of July 1, 2016:
http://umji.sjtu.edu.cn/academics/graduate-program

"Guangxi Party Representative Visits COMAC [广西党政代表团到中国商飞调研]," COMAC News Center, May 18, 2016. As of June 12, 2016:
http://www.comac.cc/xwzx/gsxw/201605/18/t20160518_3838560.shtml

Han Tianyang, "State-Owned AVIC Buys US-Based Nexteer," *China Daily*, April 11, 2011. As of July 1, 2016:
http://www.chinadaily.com.cn/business/2011-04/11/content_12306100.htm

Hanemann, Thilo, and Daniel H. Rosen, *Chinese Investment in the United States: Recent Trends and the Policy Agenda*, New York: Rhodium Group, December 2016. As of January 6, 2017:
http://origin.www.uscc.gov/sites/default/files/Research/Chinese_Investment_in_the_United_States_Rhodium.pdf

Hannas, William C., James Mulvenon, and Anna B. Puglisi, *Chinese Industrial Espionage: Technological Acquisition and Military Modernization*, New York: Routledge, 2013.

"He Dongfeng Meets Sichuan Provincial Secretary Yin Li [贺东风拜会四川省省长尹力]," COMAC News Center, April 22, 2016. As of June 12, 2016:
http://www.comac.cc/xwzx/gsxw/201604/22/t20160422_3746354.shtml

"The 'Helicopter King of China' Is Quietly Building an Empire," *Business Insider*, July 13, 2012. As of July 1, 2016:
http://www.businessinsider.com/the-helicopter-king-of-china-is-quietly-building-an-empire-2012-7

"Hong Kong NPC Research Team Visits COMAC [港区全国人大代表专题调研团走进中国商飞]," COMAC News Center, May 8, 2016. As of June 12, 2016:
http://www.comac.cc/xwzx/gsxw/201605/08/t20160508_3797070.shtml

Huber, Mark, "Epic Sold to Russian MRO," AIN online, April 2, 2012. As of August 30, 2016:
http://www.ainonline.com/aviation-news/aviation-international-news/2012-04-02/epic-sold-russian-mro

Jiang, Steven, "China to Take on Boeing, Airbus with Homegrown C919 Passenger Jet," CNN.com, November 2, 2015. As of July 18, 2016:
http://www.cnn.com/2015/11/02/asia/china-new-c919-passenger-jet

Jilin Hanxing Group, "Events [大事记]," website, undated. As of July 1, 2016:
http://www.hxjtchina.com/events.asp

———, "Group Profile," website, undated. As of October 3, 2016:
http://www.hxjtchina.com/eng/about.asp

"Joint Center for Energy Research Promotes U.S.-China Collaboration," *Penn State News*, October 21, 2014. As of July 1, 2016:
http://news.psu.edu/story/331299/2014/10/21/academics/
joint-center-energy-research-promotes-us-china-collaboration

"Joint Institute," Shanghai Jiao Tong University website, undated. As of July 1, 2016:
http://en.sjtu.edu.cn/admission/joint-institute

"Laboratories," University of Michigan-Shanghai Jiao Tong University Joint Institute website, undated. As of July 1, 2016:
http://umji.sjtu.edu.cn/research/labs

"Law of the People's Republic of China on Joint Ventures Using Chinese and Foreign Investment," China.org.cn, March 16, 2007. As of June 21, 2016:
http://www.china.org.cn/english/government/207001.htm

Lin, Jeffrey, and P. W. Singer, "China's Armed Robot Helicopter Takes Flight," *Popular Science*, July, 11, 2016. As of December 5, 2016:
http://www.popsci.com/chinas-armed-robot-helicopter-takes-flight

Lynch, Kerry, "Continental Motors Adding Avionics to Expanding Capabilities," *Aviation Week Intelligence Network*, June 24, 2014. As of December 2, 2016:
http://aviationweek.com/awin-only/
continental-motors-adding-avionics-expanding-capabilities

McCartney, Scott, "How Airlines Spend Your Airfare," *Wall Street Journal*, June 6, 2012. As of July 15, 2016:
http://www.wsj.com/articles/SB10001424052702303296604577450581396602106

"Middle- and Long-Term Development Plan for the Civil Aviation Industry (2013–2020) [国家中长期科学和技术发展规划纲要 (2012－2030年)]," Ministry of Industry and Information Technology, May 22, 2013.

"Mooney 'Hibernation' Ends, Texas Factory Is Humming," *AOPA*, May 5, 2015. As of July 1, 2016:
https://www.aopa.org/news-and-media/all-news/2015/may/pilot/f_mooney

"Mooney International Appoints New President and CEO," *Mooney*, August 16, 2016. As of August 29, 2016:
http://www.mooney.com/en/pr.html#blog22

"News," Ohio State University website, undated. As of July 1, 2016:
https://oia.osu.edu/161-gateways/china.html

"News," Sherpa Aircraft, July 22, 2016. As of August 30, 2016:
http://www.sherpaaircraft.com/news/

Nexteer Automotive, "History," website, undated. As of July 1, 2016:
http://www.nexteer.com/history/

"Ningxia Hui Minority Special Administrative Zone Vice Party Secretary Li Rui Visits COMAC [宁夏回族自治区副主席李锐到中国商飞调研]," COMAC News Center, May 7, 2016. As of June 12, 2016:
http://www.comac.cc/xwzx/gsxw/201605/07/t20160507_3794977.shtml

Office of Transportation and Machinery, Aerospace Team, Industry Reports, "China (2013)," International Trade Administration, undated. As of June 21, 2016:
http://www.trade.gov/td/otm/assets/aero/China2013.pdf

O'Neil, Brendan, Shane Norton, Leslie Levesque, Charlie Dougherty, and Vardan Genanyan, *Aerospace and Defense Economic Impact Analysis: A Report for the Aerospace Industries Association*, IHS Economics, April 2016. As of December 14, 2016:
http://www.aia-aerospace.org/wp-content/uploads/2016/05/AD_Industry_Economic_Impact_Analysis_Final.pdf

Osburg, Jan, Philip S. Anton, Frank Camm, Jeremy M. Eckhause, Jaime Hastings, Jakub Hlavka, James G. Kallimani, Thomas Light, Chad J. R. Ohlandt, Douglas Shontz, Abbie Tingstad, and Jia Xu, *Expanding Flight Research Capabilities, Needs, and Management Options for NASA's Aeronautics Research Mission Directorate*, Santa Monica, Calif.: RAND Corporation, RR-1361-NASA, 2016. As of December 7, 2016:
http://www.rand.org/pubs/research_reports/RR1361.html

"Overview," University of Maryland–China Research Park website, undated. As of July 1, 2016:
http://www.umcrp.umd.edu/overview.html

Page, Jeremy, "China Eyes U.S. Defense Contracts," *Wall Street Journal*, February 4, 2011. As of September 16, 2016:
http://www.wsj.com/articles/SB10001424052748704775604576119811508921144

"The Partnership," USC Viterbi School of Engineering website, undated. As of July 1, 2016:
http://viterbi.usc.edu/academics/globalization/research-collaborations/xidian-university.htm

"Partnership Programs," Case School of Engineering website, undated. As of July 1, 2016:
http://engineering.case.edu/partnership_degree_programs

People's Republic of China, National People's Congress, "Work Report on 10th Five-Year Plan for the Country's Economic and Social Development [中华人民共和国国民经济和社 会发展第十个五年规划纲要]," Beijing, March 2001.

———, "Work Report on 11th Five-Year Plan for the Country's Economic and Social Development [中华人民共和国国民经济和社 会发展第十一个五年规划纲要]," Beijing, March 2006.

———, "Work Report on 12th Five-Year Plan for the Country's Economic and Social Development [中华人民共和国国民经济和社 会发展第十二个五年规划纲要]," March 2011.

———, China's 13th (2015–2020) Five-Year Plan, Beijing, 2015.

———, "Work Report on 13th Five-Year Plan for the Country's Economic and Social Development [中华人民共和国国民经济和社会发展第十三个五年规划纲要]," March 17, 2016.

Perez, Bien, "Chinese Direct Investment in U.S. to Top US $10 Billion for Third Year in a Row," *South China Morning Post*, November 13, 2015. As of July 1, 2016:
http://www.scmp.com/business/companies/article/1878437/chinese-direct-investment-us-top-us10-billion-third-year-row

Perrett, Bradley, "Chinese Bizjet Mismatch: Demand vs. Assembly Plans," *Aviation Week and Space Technology*, October 14, 2013. As of July 1, 2016:
http://aviationweek.com/awin/chinese-bizjet-mismatch-demand-vs-assembly-plans

———, "Cessna Downsizes Its Chinese Assembly Plans," *Aviation Week and Space Technology*, April 7, 2014. As of August 29, 2016:
http://aviationweek.com/business-aviation/cessna-downsizes-its-chinese-assembly-plans

———, "Embraer, Avic Will Close Joint Harbin Company," *Aviation Week and Space Technology*, June 4, 2016a. As of September 9, 2016:
http://aviationweek.com/awinbizav/embraer-avic-will-close-joint-harbin-company

———, "Aero Engine Corp. of China Inaugurated, Separated from Avic," *Aviation Week and Space Technology*, September 2, 2016b. As of December 2, 2016:
http://aviationweek.com/new-civil-aircraft/aero-engine-corp-china-inaugurated-separated-avic

Phelps, Mark, "Superior Air Parts Is Back from Bankruptcy," *Flying*, August 5, 2010. As of July 1, 2016:
http://www.flyingmag.com/news/superior-air-parts-back-bankruptcy

"PKU and USC Sign i-Podium Educational Cooperation Program," Peking University website, May 26, 2009. As of July 1, 2016:
http://english.pku.edu.cn/news_events/news/global/66.htm

Planespotters, "Antonov An-148 Operators," website, undated. As of September 16, 2016:
https://www.planespotters.net/operators/Antonov/An-148

Polyzos, Christina, "Boston University and Zhejiang University Agreement," Boston University Center for Information and Systems Engineering website, November 5, 2014. As of July 1, 2016:
http://www.bu.edu/systems/2014/11/05/boston-university-and-zhejiang-university/

Pope, Stephen, "Cessna to Assemble Caravans in China," *Flying*, May 3, 2012. As of July 1, 2016:
http://www.flyingmag.com/aircraft/turboprops/cessna-assemble-caravans-china

———, "Mooney Aviation Back in Business," *Flying*, October 15, 2013. As of August 29, 2016:
http://www.flyingmag.com/aircraft/pistons/mooney-aviation-back-business

Pudong New Area Government, "Light Sport Aircraft to Be Produced in Pudong," e-Pudong, July 13, 2015. As of July 1, 2016:
http://english.pudong.gov.cn/html/pden/pden_ap_news_en/2015-07-13/Detail_74834.htm

"Purdue Inks Partnerships with Nanshan, Tsinghua on Educational, Research Opportunities," Purdue University website, July 1, 2013. As of July 1, 2016:
http://www.purdue.edu/newsroom/releases/2013/Q3/purdue-inks-partnerships-with-nanshan,-tsinghua-on-educational,-research-opportunities.html

Pyadushkin, Maxim, and Bradley Perrett, "Russo-Chinese Widebody Concept Design Underway," *Aviation Week and Space Technology*, February 11, 2015.

———, "Russia, China Agree On Long-Range, Widebody Airliner Partnership," *Aviation Week and Space Technology*, July 12, 2016.

Russian Aviation Insider, "SSJ 100 Production Rates Are Down 54%," website, October 5, 2015. As of July 15, 2016:
http://www.rusaviainsider.com/ssj-100-production-rates-are-down-54/

"Samueli School to Collaborate with China's Dalian University of Technology," UCI Samueli School of Engineering website, December 4, 2015. As of July 1, 2016:
http://engineering.uci.edu/news/2015/12/samueli-school-collaborate-china-s-dalian-university-technology

Saunders, Phillip C., and Joshua Wiseman, "China's Quest for Advanced Aviation Technologies," in Richard P. Hallion, Roger Cliff, and Phillip C. Saunders, eds., *The Chinese Air Force: Evolving Concepts, Roles, and Capabilities*, Washington, D.C.: National Defense University Press, 2012.

Schafer, Rachel Cao, "Tsinghua-Berkeley Shenzhen Institute Inaugurated in China," Berkeley Engineering website, October 26, 2015. As of July 1, 2016:
http://engineering.berkeley.edu/2015/10/tsinghua-berkeley-shenzhen-institute-inaugurated-china

"SCPKU Anchor Programs," Stanford Center at Peking University website, undated. As of July 1, 2016:
http://scpku.fsi.stanford.edu/content/scpku-anchor-programs

Seventh U.S.-China Aviation Summit, Washington, D.C., June 19–21, 2016.

"Shandong and U.S. Cooperate to Produce First Light Helicopter Next Year [山东与美国合作明年将生产出首架轻型直升机]," *Sina*, September 3, 2009. As of July 1, 2016:
http://finance.sina.com.cn/roll/20090903/08356702825.shtml

Shanghai Aircraft Customer Service Co., homepage, undated. As of June 6, 2016: http://sc.comac.cc/

Shanghai Aircraft Design and Research Institute, homepage, undated. As of June 6, 2016:
http://sadri.comac.cc/

"Shanghai Aircraft Design and Research Institute [上海飞机设计研究院成立]," Sina News, December 1, 2009. As of June 7, 2016:
http://news.sina.com.cn/o/2009-12-01/082516696953s.shtml

"Shanghai Aircraft Manufacturing Co.," COMAC Member Organizations, COMAC website, undated. As of June 6, 2016:
http://english.comac.cc/home/subsidiaryenterprises/201012/27/t20101227_410346.shtml

Shanghai Aircraft Manufacturing Company, homepage, undated. As of June 6, 2016:
http://samc.comac.cc/

Shanghai Municipal People's Government, "Shanghai Municipal Commission of Economy and Informatization," website, June 7, 2010. As of June 11, 2016:
http://www.shanghai.gov.cn/shanghai/node27118/node27386/node27400/node27836/userobject22ai38992.html

Shen Pin-Luen, "China's Aviation Industry: Past, Present and Future," in Richard P. Hallion, Roger Cliff, and Phillip C. Saunders, eds., *The Chinese Air Force: Evolving Concepts, Roles, and Capabilities*, Washington, D.C.: National Defense University Press, 2012.

"Sherpa Aircraft Deal with Chinese Investor Fails," AOPA, April 3, 2014. As of July 1, 2016:
http://www.aopa.org/News-and-Video/All-News/2014/April/03/sherpa-china-deal-fails

"Sherpa Strikes China Deal for Certification," AOPA, February 7, 2013. As of July 1, 2016:
http://www.aopa.org/News-and-Video/All-News/2013/February/07/Sherpa-strikes-China-deal-for-certification

Sideline conversation, Seventh U.S.-China Aviation Summit, Washington, D.C., June 19–21, 2016.

"Sikorsky and Changhe Sign Agreement for S-76D Cabin Production in China," Sikorsky, September 5, 2013. As of July 1, 2016:
http://www.sikorsky.com/pages/AboutSikorsky/PressreleaseDetails.aspx?pressreleaseid=68

So, Charlotte, "Chinese Consortium Cleared for ILFC Deal," *South China Morning Post*, June 1, 2013. As of July 1, 2016:
http://www.scmp.com/business/china-business/article/1250926/chinese-consortium-cleared-ilfc-deal

Spector, Mike, "Hawker Sales Talks Collapse over Review Worries," *Wall Street Journal*, October 18, 2012. As of July 1, 2016:
http://www.wsj.com/articles/SB10000872396390443684104578064402725144988

Spreen, Wesley E., *Marketing in the International Aerospace Industry*, Hampshire, UK: Ashgate Publishing, 2007.

State Council of the People's Republic of China, "PRC State Council Notice on Program for the Development of Strategic Emerging Industries During the 12th Five-Year Program Period [国务院关于印发'十二五'国家战略性新兴产业发展规划的通知]," July 9, 2012.

———, "State Council Notice on Printing Made in China 2025 [国务院关于印发《中国制造2025》的通知]," May 8, 2015.

———, "China to Boost General Aviation," May 17, 2016. As of June 29, 2016:
http://english.gov.cn/policies/latest_releases/2016/05/17/content_281475351384794.htm

State-Owned Assets Supervision and Administration Commission (SASAC), "List of Central Enterprises," SASAC website, August 3, 2016. As of September 9, 2016:
http://www.sasac.gov.cn/n86114/n86137/index.html

"Strategic Emerging Industries: A New Economic Engine for Chinese Economic Development [战略性新兴产业：中国经济发展的新引擎]," *China Economic Herald*, July 28, 2011. As of May 29, 2016:
http://www.ceh.com.cn/ceh/llpd/2011/7/28/83744.shtml

"Supplement No. 4 to Part 744—Entity List," Bureau of Industry and Security, September 7, 2016. As of December 14, 2016:
https://www.bis.doc.gov/index.php/documents/regulation-docs/691-supplement-no-4-to-part-744-entity-list/file

"Teledyne Technologies Agrees to Sell Teledyne Continental Motors to AVIC International," Teledyne Technologies, December 14, 2010. As of July 1, 2016:
http://teledyne.com/news/tdy_12142010.asp

Thurber, Matt, "Chinese Firm to Buy Epic Assets," AIN online, April 30, 2010. As of July 1, 2016:
http://www.ainonline.com/aviation-news/aviation-international-news/2010-04-30/chinese-firm-buy-epic-assets

Toh, Mavis, "Embraer Could Cease Production at Harbin," *FlightGlobal*, September 7, 2015. As of September 9, 2016:
https://www.flightglobal.com/news/articles/
embraer-could-cease-production-at-harbin-416357/

Traynor, Paul, "China Just Unveiled Its First Large Passenger Plane," Associated Press, November 2, 2015.

"U of I Signs Joint Education Agreement with Chinese University," Engineering at Illinois website, October 21, 2010. As of July 1, 2016:
http://engineering.illinois.edu/news/
article/2010-10-21-u-i-signs-joint-education-agreement-with-chinese-university

"UF Connections in China," University of Florida website, 2015. As of July 1, 2016:
https://www.ufic.ufl.edu/Documents/CountryBrief_China.pdf

"Undergraduate Program," University of Michigan-Shanghai Jiao Tong University Joint Institute website, undated. As of July 1, 2016:
http://umji.sjtu.edu.cn/academics/undergraduate-program

United States Bankruptcy Court for the Northern District of Texas, Dallas Division, "Memorandum Opinion and Order Denying Motion to Enforce for Lack of Subject Matter Jurisdiction," August 20, 2014. As of July 1, 2016:
http://www.txnb.uscourts.gov/sites/txnb/files/opinions/08-36705_memop.pdf

"University Abandons Plans to Establish Joint College in China," *Observer*, April 12, 2016. As of July 1, 2016:
http://ndsmcobserver.com/2016/04/
university-abandons-plans-establish-joint-college-china

"The University of Nottingham Signs £3 Million Deal with AVIC, One of China's Biggest Aerospace Businesses," University of Nottingham website, undated. As of July 1, 2016:
https://www.nottingham.ac.uk/fabs/beis/documents/avic-case-study-final.pdf

"U.S. Aerospace, Inc. and AVIC International Holding Corporation Enter into Strategic Cooperation Agreement," *Business Wire*, September 20, 2010. As of September 16, 2016:
http://www.businesswire.com/news/home/20100920006106/
en/U.S.-Aerospace-AVIC-International-Holding-Corporation-Enter

U.S. Chamber of Commerce, *Faces of Chinese Investment in the United States*, 2012. As of July 1, 2016:
https://www.uschamber.com/sites/default/files/legacy/
reports/16983_INTL_FacesChineseInvest_copyright_lr.pdf

U.S.-China Business Council, *USCBC China Economic Reform Scorecard: Steps Forward Undermined by Steps Back*, October 2016. As of January 21, 2017:
https://www.uschina.org/sites/default/files/USCBC%20Oct%202016%20
China%20Economic%20Reform%20Scorecard%20-%20Full%20Report_0.pdf

U.S. Congress, Office of Technology Assessment, *Competing Economies: America, Europe, and the Pacific Rim*, Washington, D.C.: U.S. Government Printing Office, OTA-ITE-498, October 1991.

U.S. Department of Commerce, Census Bureau, "U.S. International Trade Statistics, 2016." As of July 1, 2016:
http://censtats.census.gov/cgi-bin/naic3_6/naicCty.pl

"U.S. Universities Set Up Shop in China," *China Daily*, January 3, 2014. As of July 1, 2016:
http://usa.chinadaily.com.cn/epaper/2014-01/03/content_17214504.htm

"USC and Beijing Jiaotong University—Program," USC Viterbi School of Engineering website, undated. As of July 1, 2016:
http://viterbi.usc.edu/academics/globalization/jiaotong561504.htm

"USC and Tsinghua University (THU) Program," USC Viterbi School of Engineering website, undated. As of July 1, 2016:
http://viterbi.usc.edu/academics/globalization/research-collaborations/tsinghua-university.htm

"USC Viterbi to Explore Research and Education Collaborations in Shenzhen, China," USC Viterbi School of Engineering website, November 1, 2011. As of July 1, 2016:
http://viterbi.usc.edu/news/news/2011/usc-viterbi-to.htm

"U.VA. in China: Officials to Celebrate Grand Opening of Shanghai Office," *UVA Today*, March 3, 2015. As of July 1, 2016:
https://news.virginia.edu/content/
uva-china-officials-celebrate-grand-opening-shanghai-office

"UW and Tsinghua University Create Groundbreaking Partnership with Launch of the Global Innovation Exchange," *UW Today*, June 18, 2015. As of July 1, 2016:
http://www.washington.edu/news/2015/06/18/uw-and-tsinghua-university-create-groundbreaking-partnership-with-launch-of-the-global-innovation-exchange

Venere, Emily, "Purdue, China Forming Joint Energy-Research Labs," Purdue University website, summer 2011. As of July 1, 2016:
https://engineering.purdue.edu/EngineeringImpact/2011_2/Spotlights/
ChangeTheWorld/purdue-china-forming-joint-energyresearch-labs

"Viterbi School Builds Strategic Bridges with Tsinghua, China's Top Technical University," USC Viterbi School of Engineering website, June 3, 2007. As of July 1, 2016:
http://viterbi.usc.edu/news/news/2007/viterbi-school-builds.htm

Wang Jun, "COMAC's First U.S. Subsidiary Set to Take Off," *China Daily*, November 26, 2013. As of December 2, 2016:
http://www.chinadaily.com.cn/business/overseascnep/2013-11/26/
content_17131443.htm

Wang Tao, "We Should Cheer the High Expectations of the C919 [王韬: 应为C919的超预期成果喝彩]," *Observer* [观察者], November 3, 2015. As of June 20, 2016:
http://www.guancha.cn/tui-si/2015_11_03_339808_s.shtml

Whitfield, Bethany, "Cirrus Completes Merger with Chinese Firm CAIGA," *Flying*, June 29, 2011. As of July 1, 2016:
http://www.flyingmag.com/news/cirrus-completes-merger-chinese-firm-caiga

———, "Cessna Signs Deal to Build Jets, Other Aircraft in China," *Flying*, March 27, 2012. As of July 1, 2016:
http://www.flyingmag.com/news/cessna-signs-deal-build-jets-other-aircraft-china

Wood, Janice, "Embry-Riddle Partners with Chinese Universities," *General Aviation News*, November 30, 2011. As of July 1, 2016:
http://generalaviationnews.com/2011/11/30/embry-riddle-partners-with-chinese-universities

"Wuhu Mayor Pan Chaohui Visits COMAC [芜湖市市长潘朝晖到中国商飞访问]," COMAC News Center, May 9, 2016. As of June 12, 2016:
http://www.comac.cc/xwzx/gsxw/201605/09/t20160509_3801532.shtml

Yin Pumin, "C919, Made in China," *Beijing Review*, November 19, 2015. As of June 21, 2016:
http://www.bjreview.com/Nation/201511/t20151112_800042425.html

Yu Dawei [于达维], "CAIGA Acquires U.S. Plane Manufacturer [中航工业通飞并购美国飞机制造商]," Caixin Online [财新网], March 1, 2011. As of July 1, 2016:
http://companies.caixin.com/2011-03-01/100230717.html

Yu Wei, "UC-Berkeley Linked R&D Center Opens in Shanghai," *China Daily*, November 20, 2013. As of July 1, 2016:
http://usa.chinadaily.com.cn/epaper/2013-11/20/content_17118231.htm

"Yunnan Party Representative Visits COMAC [云南省党政代表团走进中国商飞]," COMAC News Center, April 20, 2016. As of June 12, 2016:
http://www.comac.cc/xwzx/gsxw/201604/20/t20160420_3737990.shtml

Zhangjiang Hi-Tech Park, "About Us," website, undated. As of June 8, 2016:
http://www.zjpark.com/cn/AboutUs.html

Zhu Qionghua [朱琼华], "'Acquisition Maniac' Cheng Shenzong: 'Neophyte' and 'Hyper' ['并购狂人'成身棕: '外行'与'炒炸者']," *Phoenix Finance* [凤凰财经], July 19, 2012. As of September 9, 2016:
http://finance.ifeng.com/news/people/20120719/6783115.shtml

Zimmerman, John, "The Skycatcher's Death Proves the LSA Rule Is a Failure," *Air Facts*, April 21, 2014. As of August 29, 2016:
http://airfactsjournal.com/2014/04/skycatchers-death-proves-lsa-rule-failure/